T0318386

Conceptual Breakthroughs in

THE EVOLUTIONARY
BIOLOGY OF AGING

Conceptual Breakthroughs

CONCEPTUAL BREAKTHROUGHS in
THE EVOLUTIONARY
BIOLOGY OF AGING

KENNETH R. ARNOLD
*Department of Ecology and Evolutionary Biology,
University of California, Irvine, CA, United States*
MICHAEL R. ROSE
*Department of Ecology and Evolutionary Biology,
University of California, Irvine, CA, United States*

Series editor

JOHN C. AVISE
*Department of Ecology and Evolutionary Biology,
University of California, Irvine, CA, United States*

ACADEMIC PRESS

An imprint of Elsevier

ELSEVIER

Academic Press is an imprint of Elsevier
125 London Wall, London EC2Y 5AS, United Kingdom
525 B Street, Suite 1650, San Diego, CA 92101, United States
50 Hampshire Street, 5th Floor, Cambridge, MA 02139, United States
The Boulevard, Langford Lane, Kidlington, Oxford OX5 1GB, United Kingdom

ISBN: 978-0-12-821545-6

For information on all Academic Press publications visit our
website at https://www.elsevier.com/books-and-journals

Publisher: Nikki P. Levy
Acquisitions Editor: Simonetta Harrison
Editorial Project Manager: Kathrine Esten
Production Project Manager: Omer Mukthar
Cover Designer: Matthew Limbert

Typeset by TNQ Technologies

Working together
to grow libraries in
developing countries

www.elsevier.com • www.bookaid.org

Dedication

To the late Marlene Arnold and the late Barry Rose.

Contents

Foreword from the Series Editor, John C. Avise

The *Conceptual Breakthroughs* (CB) series of books by Elsevier aims to provide panoramic overviews of various scientific fields by encapsulating and rating each discipline's major historical achievements in an illuminating chronological format. Each volume in the CB series is authored by one or two world-leading experts who offer their personal insights on the major conceptual breakthroughs that have propelled a field forward to its current state of understanding. Intended for advanced undergraduates, graduate students, professionals, and interested laypersons, the dozens of essays in each CB book recount how and when a recognizable discipline achieved major advances along its developmental pathway, thereby offering readers a pithy historical account of how that field came to be what it is today.

This is the fourth volume in the CB series—all written in the same concise style and format—and all intended for an intellectually curious audience ranging from laypersons and beginning students to advanced practitioners. The first three books in the CB series were as follows:

Conceptual Breakthroughs in Evolutionary Genetics by John C. Avise (2014).

Conceptual Breakthroughs in Ethology and Animal Behavior by Michael D. Breed (2017).

Conceptual Breakthroughs in Evolutionary Ecology by Laurence D. Mueller (2020).

CHAPTER ONE

Introduction

Aging is a topic like few others in biology. While scientific problems like speciation or the evolution of sex are more foundational for evolutionary biology, they have little significance in our everyday lives. The only topic of comparable practical importance is the biology of infectious disease, which impinges on our continued survival. The years since 2019 have only emphasized the value of understanding the underlying biology of infection for not only our survival, but also our happiness, and our economic productivity.

But infectious disease is not comparable to aging in our capacity to effectively mitigate its impact medically. Medical progress against the threat posed by infection has been sustained and profound, virtually miraculous compared to the toll of such disease on human lives from 1840 to 1920. This progress was due in large part to the destruction of the classical miasma theories of disease at the hands of microbiologists during that same period. And for us the contrast between the present impotence of medicine in the face of the depredations of aging and the prepotent power of medicine applied to the problem of infection is highly motivating (Rose et al., 2017).

A common reaction to this contrast among both physicians and biologists is to complain that our scientific understanding of the foundations of biological aging is not remotely comparable to our scientific understanding of infection. On that point, we emphatically dissent. Indeed, this volume amounts to an extended argument for the view that we do in fact have an excellent scientific understanding of the causal foundations of aging. Our scientific understanding of the infectious disease was developed over a protracted period. That period began in 1840, when the microbial theory of infectious disease was distinguished from miasma theories of infection (Henle, 1840), to 1920, when resistance to the microbiological analysis of infection was finally swept away by the influenza pandemic of 1918–20 (Johnson & Mueller, 2002; Barro et al., 2020). For the study of aging, our view is that key breakthroughs were made over a period from 1941 to 2020, as we shall document in this book.

In terms of basic science, we will argue that we now have foundations for understanding and mitigating aging in a manner comparable to the chemists,

Conceptual Breakthroughs in The Evolutionary Biology of Aging
ISBN: 978-0-12-821545-6
https://doi.org/10.1016/B978-0-12-821545-6.00063-7

biologists, and physicians of the 1920s who were at the forefront of medical microbiology. Yet these foundations have not been accepted across the full range of biology, while the medical treatment of aging is as haphazard and ineffectual as the medical treatment of infection was before the end of the 19th Century. If we are correct, this situation is a travesty that imperils the survival and health of hundreds of millions of older people.

We believe that we can show the interested and objective reader why this troubling situation exists. Simply put, reductionist cell-molecular biologists have claimed the biological problem of aging as their domain for scientific research, much like the miasmaticists of the 19th Century claimed that their research on "bad air" was the key to understanding and treating infection. In staking out the problem of aging as part of their disciplinary fiefdom, we believe that cell-molecular biologists have sustained scientific confusion about the problem. Worse, we believe that they have set back geriatrics by 50 years, if not more.

At its root, the problem is that few appreciate the research successes that spring from the evolutionary theory of aging. The ignorance of this body of research results in both academic and popular presses replete with articles and books that bemoan the inadequacy of our understanding of aging. Worse, there are other articles and books by reductionist authors which claim that *they have* achieved their own recent breakthroughs in the conceptual foundations of aging, promising a bright future sure to lead to the easy and rapid abolition of aging (e.g., de Grey and Rae, 2008; Sinclair, 2019). With such grandiose claims, the latter type of publication only invites skepticism as to whether or not we understand biological aging at all.

Our opinions are rooted in evolutionary research on aging. Like many evolutionary biologists, we believe that the research and publications of our close colleagues have provided strong and indispensable foundations for the understanding and eventual treatment of aging. Furthermore, over the course of the last 80 years, evolutionary biologists have repeatedly falsified the claims of cell-molecular biologists. Again in our opinion, this body of work is the greatest achievement of evolutionary biology over the last 80 years, when viewed from the standpoint of potential significance for the mitigation of human suffering.

Here we review the breakthroughs that have built the scientifically powerful field that is the evolutionary biology of aging. These breakthroughs have sometimes been "merely" conceptual. But more often than not the field's conceptual breakthroughs have been derived from and developed in conjunction with careful comparative research, explicit mathematical analysis, and strong-inference laboratory experiments.

It is no small part of our task, as authors, to point out in detail the many cases in which the common assumptions of cell-molecular research on aging have been obliterated by comparative and experimental findings in the evolutionary biology of aging. Know that when we point out such errors our intent is to clear away underbrush. We are well aware of the traditional practice in academic biology to avoid pointing out the errors of well-established views. Our view is that this tradition is a pernicious practice that is sustained by the power of anonymous reviewers who would otherwise face their diminution. But it is only through falsification that a multiplicity of contending theories can be winnowed.

Let us now turn to an outline of our narrative. The evolutionary biology of aging is a subfield within evolutionary biology as a whole. As such, its foundations and its origins come from evolutionary biology itself. Its explicit primordia are to be found in late 19th-century writings of August Weismann, theoretical population genetics developed by Norton (1928), and some groundbreaking Drosophila experiments from Raymond Pearl and his students starting in the 1920s (e.g., Pearl, 1922). Right from the earliest speculations of Weismann, the evolutionary approach to aging was a clear break from the widespread, and continuing, presupposition that the causes of aging are chiefly or merely physiological, the latter being an idea first delineated by Aristotle millennia ago.

The evolutionary theory of aging didn't really coalesce until after the landmark paper by William D. Hamilton, published in 1966, *The moulding of senescence by natural selection*. During the 1970s, the field was chiefly advanced by Brian Charlesworth's integration of Hamilton's findings into the theoretical population genetics tradition founded by Norton (1928), culminating in the definitive Charlesworth (1980) monograph, *Evolution in Age-Structured Populations*. The 1980s then provided sustained experimental research on the evolutionary biology of aging, much of it summarized and connected to prior theory in the book *Evolutionary Biology of Aging* (Rose, 1991).

Our aim for this installment in the Conceptual Breakthroughs series is to take stock of the major shifts in research on the evolutionary biology of aging, to clarify the many misconceptions still prevalent in gerontology, and to establish the indispensable role of evolutionary findings for aging research generally. In doing so, we attempt a sustained case for the cogency, validity, and utility of the evolutionary biology of aging, as opposed to the merely physiological explanation, manipulation, or analysis of aging.

Each chapter highlights a specific *conceptual breakthrough* in the development of the evolutionary biology of aging. These breakthroughs are embodied primarily by landmark articles that furthered the understanding of the evolution of aging by providing new insights or findings. To illustrate such changes in course, each adduced breakthrough is preceded by a description of the relevant *standard paradigm* which was discarded, revised, or reformed in light of the conceptual breakthrough. Many of our chapters feature the tension between standard physiological explanations of aging versus contrasting evolutionary theories or experiments. In sum, we believe that the overall arc of this book amounts to a devastating case against strictly nonevolutionary approaches to aging, as well as a case for the centrality of evolutionary biology in the study of aging, from its physiological mechanisms to its evolutionary determinants.

In addition, the conceptual breakthrough discussed in each chapter is assigned a score (1—10) that estimates the relative merits of each work in a concluding section; denoted as the *Impact* score. This format has been standard in the Conceptual Breakthroughs series from its first volume (Avise, 2014). High impact scores indicate conceptual breakthroughs of great importance for the field, whether through their novelty, powerful union of theory with experimentation, pragmatic implications, or overall relevance. Relatively lower ranking research is merely less essential to the field covered by this particular installment of the series, the evolutionary biology of aging, whatever their other merits. In addition, we can't claim omniscience, so our scorings are by no means definitive. Perhaps the true value of our scores will be to spark discussion among academics with an interest in the field, especially in seminar courses or in departmental break rooms.

Over the last 30 years, the evolutionary biology of aging has made substantive and widely recognized contributions across the spectrum of subdisciplines within biology that are founded on Darwinian reasoning. It is our view that the non-Darwinian fields which discuss aging still have much to be gained from the evolutionary biology of aging, especially mainstream biological gerontology. Indeed, we contend that this book amounts to a case that a paradigm shift toward evolutionary reasoning is necessary for progress to be made in aging research.

In this installment of the Conceptual Breakthroughs series, we provide a comprehensive outline of the major accomplishments made in evolutionary research on the topic of aging, which should spur no serious controversies among evolutionists save for any perceived omissions due largely to limitations in the length of this book. More controversial perhaps is our goal of prompting modern gerontologists to consider an evolutionary framework when conducting their research, instead of a merely reductionist cell-molecular perspective. [But note that we are not disputing the importance of cell-molecular mechanisms in or of themselves, providing that they are correctly and robustly identified.] This more expansive agenda for gerontology, as we view it, does not call for a complete overhaul of current methods. Rather, we hope for a greater appreciation of the core themes outlined in this volume, including the importance of evolutionary demography, age-specificity, replication, and genetic diversity for aging research.

The present volume, in short, is a straightforward update and commentary on the evolutionary biology of aging for those who are aspiring or current evolutionary geneticists, evolutionary physiologists, and the like. As such, it will be a useful resource for advanced courses in those fields. However, we hope that this book will *also* serve as a stimulus for discussion and debate beyond the Darwinian wing of biology. From either standpoint, we believe that we have supplied a step-by-step, study-by-study introduction to the evolutionary biology of aging. We hope that many of those who work on the problems of aging, or who study its literature, will find it a useful resource.

Lastly, we would like to thank John Avise for the opportunity to contribute to the Conceptual Breakthrough series, as well as thank Laurence Mueller, Zachary Greenspan, Robert Shmookler Reis, Molly Burke, Mark Phillips, James Kezos, Joseph Graves, Valeria Chavarin, Ryan Robinson, Vinalon Mones, Grant Rutledge, David Bahry, Christian Vu, and Parvin Shahrestani, for their editorial suggestions and discussions. We are grateful for the vast amount of work done by our postdoctoral, graduate, and undergraduate students, which have produced many of the findings that we discuss here. The emotional and material support of our loving families has been indispensable. Finally, we thank the millions of flies, nematodes, and rodents who have given their lives "doing time on the inside." Only through their sacrifice was the research of our lab and the labs of our many outstanding colleagues possible.

References and further reading

Avise, J. C. (2014). *Conceptual breakthroughs in evolutionary genetics: A brief history of shifting paradigms*. Elsevier, Academic Press.

Barro, R., Ursúa, J., & Weng, J. (2020). *The coronavirus and the great influenza pandemic: Lessons from the "Spanish Flu" for the coronavirus's potential effects on mortality and economic activity.*

National Bureau of Economic Research. https://doi.org/10.3386/w26866 (No. w26866; p. w26866).

Charlesworth, B. (1980). *Evolution in age-structured populations.* Cambridge, U.K: Cambridge University Press.

De Grey, A. D. N. J., & Rae, M. (2008). *Ending aging: The rejuvenation breakthroughs that could reverse human aging in our lifetime.* St. Martin's Griffin.

Hamilton, W. D. (1966). The moulding of senescence by natural selection. *Journal of Theoretical Biology, 12*(1), 12–45. https://doi.org/10.1016/0022-5193(66)90184-6

Henle, Jakob (1840). Pathologische un*tersuchungen (1.0) [computer software].* Berlin: NN. https://doi.org/10.11588/DIGLIT.15175

Johnson, N. P., & Mueller, J. (2002). Updating the accounts: Global mortality of the 1918–1920 "Spanish" influenza pandemic. *Bulletin of the History of Medicine,* 105–115.

Norton, H. T. J. (1928). Natural selection and mendelian variation. *Proceedings of the London Mathematical Society, s2–28*(1), 1–45. https://doi.org/10.1112/plms/s2-28.1.1

Pearl, R. (1922). *The biology of death.* Philadelphia and London: J. B. lippincott Company. http://archive.org/details/biologyofdeath004218mbp.

Rose, M. R. (1991). *Evolutionary biology of aging.* Oxford University Press.

Rose, M. R., Rutledge, G. A., Cabral, L. G., Greer, L. F., Canfield, A. L., & Cervantes, B. G. (2017). Evolution and the future of medicine. In *On human nature* (pp. 695–705). Elsevier. https://doi.org/10.1016/B978-0-12-420190-3.00042-9

Sinclair, D. A., & LaPlante, M. D. (2019). *Lifespan: Why we age–and why we don't have to* (First Atria Books hardcover edition). Atria Books.

CHAPTER TWO

384–322 B.C: The first biologist on aging

The standard paradigm

From the inception of civilization, the extension of the human lifespan has been a common aspiration. As early as the legend of Gilgamesh to Taoist theories and practices, attempts to overcome human aging are widespread (Gruman, 1966). Assumptions about the cause of aging ranged widely between cultures and were often couched in spiritual or divine terms. Such literally magical thinking fostered hopes for mythical places, practices, or substances that might provide eternal youth (Gruman, 1966; Haycock, 2008). Such myths offered a myriad of remedies ranging from the palliative to the cosmetic, and even the psychological. A notable shortcoming common to all such ancient myths about aging and its amelioration is that they lacked any kind of scientific foundation.

The conceptual breakthrough

Aristotle was the first great biologist, providing foundational elements for biological science like the concepts of adaptation, species, and hierarchies of causation (Leroi, 2014). A failure to appreciate this historical fact has been commonplace among biologists ever since Darwin. Some biologists attribute Aristotle's many conceptual breakthroughs in biology to Charles Darwin, especially—we are sad to say—evolutionary biologists. This feat of cultural amnesia by biologists has been so debilitating that many academic biologists don't understand why Aristotle should even be discussed in historical narratives of biology.

Of greatest relevance for the present book, Aristotle provided a cogent scientific analysis of aging among living things, the first such analysis known to history. Tragically, his analysis would set the field of gerontology off on a course that would lead to its sustained misdirection, if not indeed virtual futility.

In his works *On Youth and Old Age, On Life and Death,* and *On Length and Shortness of Life,* Aristotle poses questions regarding the universality of aging,

Conceptual Breakthroughs in The Evolutionary Biology of Aging
ISBN: 978-0-12-821545-6
https://doi.org/10.1016/B978-0-12-821545-6.00010-8

whether it is "a single or diverse cause that makes some organisms to be long-lived" while others short-lived (Aristotle, 1957a, 1957b, 1957c). In these works, Aristotle grapples first with an explanation of death and why health is lost at different rates and in different ways among species.

Aristotle's theorizing about aging emphasized progressive and cumulative disruption of physiological functions with biological time. Like other classical theorists of medicine, he believed that life was maintained by a "vital heat" sustained by the consumption of nutrients. The organ systems responsible for sustaining and tempering this heat, he supposed, would progressively fail over time, leaving older individuals "cold and dry"; death must then come with the final extinguishing of the vital heat. Almost all subsequent theories of biological aging have simply updated the physiological particulars of this kind of cumulative process of deterioration. But the underlying assumption that aging is neither more nor less than such a cumulative physiological process is ultimately attributable to Aristotle.

Impact: 6

While useful in diverting the discussion of aging away from mysticism toward natural history and biological science, Aristotle was limited by the biological knowledge of his time. Through no fault of his own, Aristotle provided an appealing and intuitive analysis of biological aging, but one which has misled the field of gerontology ever since. His central error, which has been perpetuated for more than two millennia, was the inference that aging is no more and no less than a physiological process of deterioration. With this assumption achieving the status of an unexamined axiom, it has only been natural for physicians and biologists right up to the present time to look for entirely "mechanistic" (meaning merely physiological) explanations and analyses of aging.

This error was sustained in no small part because of the widespread belief that aging is universal among all multicellular forms of life, if not indeed *all* forms of life. When this mistaken comparative hypothesis is combined with the notion that aging is a merely physiological process, you have a combination of intuitive plausibility with a lack of imagination that has stultified aging research over millennia. As we will show, this was an intellectual disaster that has hampered the nonevolutionary approach to aging all the way to the present-day.

As evolutionary biologists, it is natural for us to wonder why Aristotle supposed something as pernicious as aging should exist at all, given that he had no notion of physical constraints like increasing entropy in closed systems. There is where some of Aristotle's deepest assumptions about the living world come into play. For Aristotle, the natural world was explained best in terms of teleological arguments, where animals are suited to their environments thanks to constituent organs operating according to understandable mechanisms. In the context provided by Aristotle, aging is tantamount to its physiological processes, such as the progressive loss of the "essential heat."

However, how this loss of functionality came to be is not wholly addressed in Aristotle's writings. As evolutionary biologists, we ask what is the purpose of aging or even death, for Aristotle? This puzzle is especially anomalous, given Aristotle's emphasis on the degree to which the anatomical and physiological characteristics of animals reflect adaptation to the environments in which they live (Leroi, 2014). Surely a pan-adaptationist must find the mere existence of aging a virtual refutation of that adaptational view of life?

Thus, for the evolutionary biology of aging, the greatest failure of Aristotle was not his crude classical explanation of *how* aging occurs, but rather *why* it exists at all. His lack of thought on this question set a precedent that has been widely sustained among gerontologists who operate outside of the Darwinian tradition.

References and further reading

Aristotle. (1957a). Parva naturalia. In W. S. Hett, Trans (Ed.), *On length and shortness of life*. Harvard University Press. https://doi.org/10.4159/DLCL.aristotle-parva_naturalia_length_shortness_life.1957 [Data set].

Aristotle. (1957b). Parva naturalia. In W. S. Hett, Trans (Ed.), *On youth and old age. On life and death*. Harvard University Press. https://doi.org/10.4159/DLCL.aristotle-parva_naturalia_youth_old_age_life_death.1957 [Data set].

Aristotle. (1957c). Parva naturalia. *On the Soul*. In W. S. Hett, Trans (Ed.), *Digital loeb classical library*. Harvard University Press. https://doi.org/10.4159/dlcl.aristotle-parva_naturalia_length_shortness_life.1957 [Data set].

Gruman, G. J. (1966). A History of Ideas about the Prolongation of life, the evolution of progenvity hypotheses to 1800. *Transactions of the American Philosophical Society, 56*(9).

Haycock, D. B. (2008). *Mortal Coil, A short history of living longer*. New Haven, Conn: Yale University Press.

Leroi, A. M., & Koutsogiannopoulos, D. (2014). *The lagoon: How Aristotle invented science*. In S. MacPherson (Ed.) (Trans.). Viking Pengiun.

Ross, G. R. T. (1908). *Translation of Aristotle's de Longitudine et brevitate vitae*. Oxford: Clarendon Press.

Woodcox, A. (2018). *Aristotle's theory of aging* (pp. 65–78). LV: Cahiers des études autochtones.

CHAPTER THREE

1645: A tale of two Bacons

The standard paradigm
Aging treatment without empiricism

Following the fall of Rome, science and medicine began to stagnate in the West. However, to the East in Byzantium, ecclesiastical scholars preserved and perpetuated knowledge of the classics, including but not limited to the works of Plato and Aristotle, and thereby sowed the seeds for a rediscovery of such work in Western Europe during the Renaissance (Markus, 1961).

One Western scholar who delved into the classical literature was the medieval philosopher Roger Bacon (1220—92), known also as Doctor Mirabilis, owing to the fantastical tales and myths that surrounded him (Power, 2006). Behind the fog of mysticism is what some believe to be an early scientist (Kuhn, 1976; Schramm, 1998). A Franciscan friar and teacher of the works of Aristotle, Roger Bacon provided commentary on an array of subjects ranging from physics to linguistics, and even to the less commonly discussed topic of aging. The precise impact Bacon had on these fields is widely debated to this day (Hackett, 1997; Mantovani, 2020). What is more commonly accepted is Roger Bacon's influence as an Aristotelian realist and an advocate of knowledge derived from sensory experience and experimentation (Hochburg, 1953). Roger Bacon's role was chiefly that of an advocate, falling well short of Aristotle's standards as a practitioner of empiricism.

On the topic of aging Roger Bacon's contributions were also limited. In his account *The Cure of Old Age, and Preservation of Youth*, Bacon employs the Aristotelian definition of aging as a merely physiological process, invoking the Greek elements of fire, water, et cetera to explain its phenomena. He describes a variety of treatments aimed at the preservation of "heat" and "moisture" to sustain one's youthfulness. These remedies are a collection of suggestions with fanciful antiaging properties like eating the "bowels of a long-lived creature," citing ancient sources without any tests of their efficacy. Finally printed and distributed in 1683, Bacon's treatise provided a

Conceptual Breakthroughs in The Evolutionary Biology of Aging
ISBN: 978-0-12-821545-6
https://doi.org/10.1016/B978-0-12-821545-6.00051-0

comprehensive account of ailments, pseudo-treatments, as well as benign advice like moderation in diet, exercise, and thought.

The conceptual breakthrough

Francis Bacon was a statesman and naturalist regarded by many to be the father of empiricism (Hochberg, 1953). In his work *Novum Organum,* Francis advocated skepticism in collecting scientific information. He proposed inductive development of theories, the first attempt at a systematic scientific method. Such a method would allow for greater stringency in testing the validity of claims about data, he argued, both of which were lacking in the compiled works of Roger Bacon.

Bacon et al. (1857) views on aging are best illustrated by the 1645 work *Historia Vitae et Mortis.* Using the same Aristotelian assumption that aging is a merely physiological process, *Historia Vitae* chronicles the origins and operations of life as it was then understood in terms of the preservation of vital heat. Though there is some overlap with Roger Bacon's work, a notable difference is the book's relative skepticism. In his similarly titled segment, *The length and shortness of Life in living creatures,* Bacon begins with the oft quoted statement that "observations are light and fabulous" in work on aging [Here the term "fabulous" refers to derivation from fables]. Likewise, in the segment *Medicines for Long Life,* he notes that while there are many "for preserving health, there are but few to prolong life." He blames "superstitious fables, and strange delusions" for failures to extend life. While many of the remedies that Francis Bacon discusses bear some resemblance to those of Roger Bacon, Francis Bacon's restraint and skepticism in his book marks a turn toward a scientific approach to the topic of aging.

Impact: 4

Between Aristotle's first discussions of aging and the cusp of the modern era, the topic of the biology of aging lacked the attention accorded to other parts of biology. While many remedies were proposed for the scourge of aging, none were based on significant scientific findings. The breakthrough provided by Francis Bacon chiefly was emphasizing scientific skepticism concerning the alleged "treatments" of his time. Like Aristotle, for both Roger and Francis Bacon their publications on the topic of aging were addenda to expansive careers. While neither Bacon did much to further our concrete understanding of aging, both championed the tools of empiricism that would be essential for the advancement of science in general. Given their tangential impact on our understanding of the evolution of aging, and their unexamined acceptance of aging as a merely physiological process, the score of four is awarded.

By this point in the development of gerontology, the assumption that aging was a merely physiological process that required no additional explanation had hardened. It would not be challenged meaningfully for more than 200 years, as we will discuss.

References and further reading

Bacon, F. (1645). Historia Vitae et Mortis. *Dillingen.*

Bacon, R. (1683). The cure of old age, and preservation of youth. *London, Printed for Tho. Flesher and Edward Evets.*

Bacon, F., (1898), *Novum Organum or true suggestions for the interpretation of nature.* London and New York.

Bacon, F., Ellis, R. L., Heath, D. D., & Spedding, J. (1857). *The works of Francis Bacon ... Collected.* In James Spedding, R. L. Ellis, & D. D. Heath (Eds.) (vol 14). London: Longman & Co. etc.

Hackett, J. (1997). *Roger Bacon and the sciences: Commemorative essays.* Leiden New York: Brill.

Hochberg, H. (1953). The empirical philosophy of Roger and Francis Bacon. *Philosophy of Science, 20*(4), 313−326. Retrieved December 14, 2020, from http://www.jstor.org/stable/185035.

Kuhn, T. S. (1976). Mathematical vs. experimental traditions in the development of physical science. *Journal of Interdisciplinary History, VII-1,* 1−31 (Summer, 1976).

Maloney, T. S. (1985). The extreme realism of Roger Bacon. *Review of Metaphysics, 38*(June), 807−837.

Mantovani, M. (2020). Visio per sillogismum. Sensation and cognition in 13th century theories of vision. In Elena Baituta (Ed.), *Medieval perceptual puzzles: Theories of perception in the 13th and 14th centuries* (pp. 111−152). Leiden: Brill.

Markus, R. A. (1961). The impact of Aristotle on medieval thought. *Blackfriars, 42*(490), 96−102 (JSTOR).

Power, A. (2006). A mirror for every age: The reputation of Roger Bacon. *The English Historical Review, 121*(492), 657−692.

Schramm, M. (1998). Experiment in Altertum und Mittelalter. In, *Vol 3. Experimental Essays—Verusche zum experiment, Zif—Interdisziplinäre Studien* (pp. 34−67). Baden-Baden (in German).

CHAPTER FOUR

1881: Natural selection is the ultimate determinant of aging

The standard paradigm

This book is not a history of the entire field of research on biological aging, commonly called gerontology. Nonetheless, the evolutionary biology of aging has developed by way of contradistinction relative to the broad field of gerontology. Thus, we will be contrasting the assumptions, concepts, and findings of the evolutionary biology of aging relative to those of conventional gerontology.

There have been recent histories of conventional gerontology (e.g., Haycock, 2008) of varying quality and detail, as well as classic works by gerontologists themselves that are somewhat historical, such as Hayflick (1979) and Finch (1994). We find them generally inadequate with respect to the clear delineation of the unspoken assumptions that animate the field of mainstream gerontology. Thus it will be part of our task to extract the key assumptions and conclusions of that genre of gerontology, like gemstones from a rocky matrix.

The most important assumption of conventional gerontology, its animating heart as it were, is Aristotle's original idea that aging is nothing more or less than one or more physiological processes that lead to cumulative failure of function over the life of an organism [Just when that process is supposed to begin will be addressed repeatedly in later chapters]. Whether such processes involve disharmony or damage, perhaps a combination of the two, has been discussed in conventional gerontology in recent decades. But almost never seriously challenged is the idea that aging is merely physiological, except by evolutionary biologists.

Here, then, is a very brief statement of this central assumption of conventional biogerontology:

Aging is a physiological process of breakdown that derives from varying forms of unavoidable physical or chemical limitations inherent to life.

Thus much of gerontology has been preoccupied with finding which physiological processes are the necessary and sufficient drivers of the degradation

Conceptual Breakthroughs in The Evolutionary Biology of Aging
ISBN: 978-0-12-821545-6
https://doi.org/10.1016/B978-0-12-821545-6.00002-9

of somatic bodies with the passing of biological time. Unexamined for millennia was the question of *why* the aging process took place at all, at least until the late nineteenth century. But in the wake of the Darwinian revolution, evolutionary biologists set about examining just such "why" questions.

The conceptual breakthrough

German evolutionary biologist August Weismann proposed that natural selection is responsible for shaping the biological longevity of an organism. Rejecting the view that an organism's physical constitution alone is responsible for senescence, Weismann proposed instead that "duration of life is really dependent upon adaptation to external conditions" whereby the length or brevity of an organism's lifespan fits the needs of species through environmental circumstance (Weismann, 1889, p. 9). Aristotle himself was fully capable of taking this view, given he was already quite interested in this kind of adaptation, long before Darwin explained the origins of adaptation in terms of natural selection. However intriguing this alternate history may be, it was not the case.

The specific selection hypothesis for the evolution of aging that Weismann first advocated was selection for the removal of older members of a population so that they would not be a detriment to the young. The adaptive presupposition appears to be that, were it not for aging, populations would be laden with infirm older individuals. Evidently relying on a type of group selection, wherein the needs of the species outweigh the needs of the individual (Williams, 1966), this theory was later abandoned even by Weismann, although his later reasoning was obscure (Kirkwood and Cremer, 1982) and of negligible subsequent influence.

Leaving aside the specifics of Weismann's group selection hypothesis, the most important innovation in Weismann's thought is the proposal that *natural selection can eliminate aging whenever it bothers to do so.* Weismann might have been directed toward this view by his awareness of the indefinite propagation of animal germ cell lineages, but we don't know if this inference is explicitly present in any of his writings. In any case, Weismann's basic proposal is supported by the existence of apparently nonaging organisms, such as fissile sea anemones, budding Hydra, creosote bushes, and trembling aspen stands (Comfort, 1979; Finch, 1994; Martinez, 1998). The contrast between these nonaging organisms and aging organisms demonstrates that merely physiological hypotheses cannot explain aging (Bell, 1984), in support of Weismann's basic suggestion.

Impact: 7

Weismann's focus on the evolution of aging arising from natural selection, as opposed to merely physiological explanations of aging, marks an important shift toward the evolutionary analysis of aging. While Weismann did compare life-history data, noting trade-offs between longevity and reproduction, he did not perform any empirical tests of his core presupposition that natural selection is the ultimate determinant of patterns of aging. Prolific in his intuition that natural selection is prepotent with respect to aging and the cost of somatic immortality, Weismann was limited by his lack of both relevant empirical data and a mathematical framework for analyzing the evolution of aging.

In the decades to follow, Weismann fell out of favor, and his contributions to the field of aging were rarely discussed. Evolutionary biologists did not follow up on his core proposals concerning aging, while mainstream gerontologists ignored his challenge to their central assumption that there is some kind of physiological necessity to aging. The effective immortality of the germ line was largely ignored as a puzzle facing mainstream gerontology.

References and further reading

Bell, G. (1984). Evolutionary and nonevolutionary theories of senescence. *The American Naturalist, 124*(4), 600—603.

Bell, G. (1988). *Sex and death in protozoa: The history of an obsession.* Cambridge University Press.

Comfort, A. (1979). *The biology of senescence* (3rd ed.). Edinburgh and London: Churchill Livingstone.

Finch, C. E. (1994). *Longevity, senescence, and the genome.* United Kingdom: University of Chicago Press.

Haycock, D. B. (2008). *Mortal Coil, A short history of living longer.* New Haven, Conn: Yale University Press.

Hayflick, L. (1979). The cell biology of aging. *Journal of Investigative Dermatology, 73*(1), 8—14. https://doi.org/10.1111/1523-1747.ep12532752

Kirkwood, T. B., & Cremer, T. (1982). Cytogerontology since 1881: A reappraisal of August Weismann and a review of modern progress. *Human Genetics, 60*(2), 101—121. https://doi.org/10.1007/bf00569695

Martínez, D. E. (1998). Mortality patterns suggest lack of senescence in Hydra. *Experimental Gerontology, 33*(3), 217—225. https://doi.org/10.1016/S0531-5565(97)00113-7

Maynard Smith, J. (1976). Group selection in predator-prey communities. *M. E. Gilpin. The Quarterly Review of Biology, 51*(2), 277—283. https://doi.org/10.1086/409311

Weismann, A., Poulton, E. B., Schönland, S., Selmar), Shipley, A. E., & Arthur, E. (1891). *Essays upon heredity and kindred biological problems.* Oxford: Clarendon press. http://archive.org/details/essaysuponheredi189101weis

Weismann, A. (1889). *Essays upon heredity and kindred biological problems.* Oxford: Claredon Press.

Williams, G. C. (1966). Natural selection, the costs of reproduction, and a refinement of lack's principle. *The American Naturalist, 100*(916), 687—690.

CHAPTER FIVE

1922: Early laboratory experiments on demography

The standard paradigm

Much of our own scientific work has been experimental, so we are sensitive to the fact that histories of scientific fields tend to focus on the great theorists. Thus, we hear much more about Newton than Tycho Brahe in histories of physics. In biology, we are told about Charles Darwin rather than Georges-Louis Leclerc, Comte de Buffon, the great 18th-century zoologist whose work on animal diversity set the stage for the evolutionists of the 19th century.

Central to the development of both mainstream gerontology and the evolutionary biology of aging has been the close study of aging in laboratory animals. This kind of work was broadly neglected before the nineteen-teens. Instead, scientific publications on aging focused on haphazard clinical data on humans, crude actuarial information from zoos, inferences from natural history reports, and the claims of agriculturalists.

Weismann provided the impetus to approach aging from an evolutionary perspective but did not himself collect empirical data. Weismann was known to compile life history data and records collected by others, in the same fashion as mainstream gerontologists. Thus strong empirical foundations for the study of the evolutionary biology of aging were lacking.

The conceptual breakthrough

Raymond Pearl is regarded as an early founder of both experimental population biology and the field of gerontology. Early work performed by Pearl and Reed (1920) modeled population growth using the logistic equation, which used a cohort's initial rate of growth r and its eventual approach to a carrying capacity K (See Fig. 5.1).

Picking up the pieces from a mouse lab lost in a fire, Pearl took the advice of T.H. Morgan to start anew with *Drosophila melanogaster,* the common laboratory fruit fly. This species provided discrete developmental stages

Conceptual Breakthroughs in The Evolutionary Biology of Aging
ISBN: 978-0-12-821545-6
https://doi.org/10.1016/B978-0-12-821545-6.00015-7

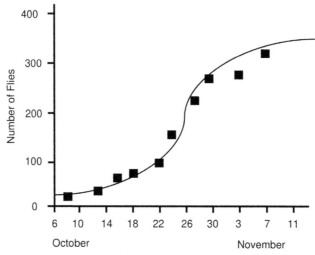

Figure 5.1 Observed population growth of *Drosophila melanogaster* relative to a fitted logistic curve. Squares represent the observed census on the given date. *Derived from Pearl, R., Miner, J. R., and Parker, S. L. (1927). Experimental studies on the duration of life. XI. Density of population and life duration in Drosophila.* The American Naturalist, 61*(675), 289–318.*

by which life tables might be constructed and large homogenous populations may be maintained.

Experiments published in Pearl's 1922 book, *The Biology of Death,* and the *American Naturalist* (Pearl & Parker, 1922) explored the net reproductive capacity of Drosophila flies in the laboratory For example, progeny were counted and recorded by sex relative to the size of the total population over a 16-day period. The observed pattern was that increasing the density of the population in a limited space resulted in a decline in the rate of reproduction per mated female. This decline in progeny production was first rapid, and then it plateaued as the population census increased in duration. Pearl concluded that the maximization of the "yield" of adult progeny depends on limits to the space and resources provided, and once those limits are exceeded, additional flies reduce the population's net reproduction. This is now a generally understood concept of density-dependence that is fundamental in ecology (vid. Mueller, 2020).

Impact: 6

Raymond Pearl starts the experimental study of the biological foundations of life-history, including the components of fitness from an age-structured perspective. Pearl shifted the study of populations from the field to the lab. This transition was not without its difficulties, however, as it is worth noting that Pearl maintained his populations in a haphazard fashion, as detailed in Sang (1949). Nevertheless, Pearl's reliance on mathematical models for explaining the simple logistic growth patterns of his populations was an early example of the use of formal models to describe experimental findings. Ultimately it is this unification of laboratory experimentation and numerical analysis that is perhaps Pearl's greatest legacy. In particular, his focus on the quantitative demography of laboratory cohorts provided a precedent that influenced subsequent research in gerontology. For the evolutionary biology of aging, such quantitative demographic data would prove indispensable for the development of the field, because its theories would be developed in terms of just such demographic parameters.

References and further reading

Mueller, L. D. (2020). *Conceptual breakthroughs in evolutionary ecology*. Academic Press, an imprint of Elsevier.

Pearl, R. (1922). *The biology of death*. Philadelphia And London: J. B. Lippincott Company. http://archive.org/details/biologyofdeath004218mbp.

Pearl, R., Miner, J. R., & Parker, S. L. (1927). Experimental studies on the duration of life. XI. Density of population and life duration in Drosophila. *The American Naturalist, 61*(675), 289–318.

Pearl, R. (1928). The rate of living, being an account of some experimental studies on the biology of life duration. New York, NY: Alfred A. Knopf. 1928.

Pearl, R., & Parker, S. L. (1922). Experimental studies on the duration of life. IV. Data on the influence of density of population on duration of life in Drosophila. *The American Naturalist, 56*(645), 312–321.

Pearl, R., & Reed, L. J. (1920). On the rate of growth of the population of the United States since 1790 and its mathematical representation. *Proceedings of the National Academy of Sciences, 6*(6), 275–288. https://doi.org/10.1073/pnas.6.6.275

Sang, J. H. (1949). Population growth in Drosophila cultures. *Biological Review, 25*, 188–219.

CHAPTER SIX

1928: Basic mathematics of selection with age-structure

The standard paradigm

Weismann was the first to approach aging from an evolutionary perspective, but he lacked the necessary mathematical framework to integrate natural selection with population genetics. While Weismann proposed the possibility of calculating fitness probabilities with age-specificity (Weismann, 1889, p. 156), he did not publish work pursuing this suggestion. Indeed, to read Weismann and other evolutionary biologists who attempted to reason about the evolution of aging using words only is a cumbersome experience, in light of the centrality of formal mathematical theory for understanding the evolution of aging.

In this respect, the evolutionary biologists who discussed aging were no different from mainstream gerontologists who followed the prevailing practice of "theoretical biologists" before the 1920s. Reasoning about such a complex quantitative phenomenon as aging using words alone is a fool's errand. Fortunately, theoretical evolutionary genetics began to develop useful mathematical tools in the years after World War I.

The conceptual breakthrough

The epoch of population genetics began with the rediscovery of Mendelian inheritance at the turn of the 20th century and gained momentum with R.A. Fisher's article *The Correlation between Relatives on the Supposition of Mendelian Inheritance* (1918). With his infinitesimal model combining Mendelian inheritance with continuous variation for phenotypic traits, Fisher established a deep statistical connection between quantitative characters and Mendelian genes. Theoretical population genetics then began to develop a strong mathematical tradition, first in the hands of Fisher, but later chiefly in the hands of J.B.S. Haldane. Many fundamental questions in the theory of natural selection acting on genetic variation were quickly resolved with the aid of well-formulated mathematical models, especially in cases with discrete generations and large population sizes (e.g., Haldane, 1932).

Conceptual Breakthroughs in The Evolutionary Biology of Aging
ISBN: 978-0-12-821545-6
https://doi.org/10.1016/B978-0-12-821545-6.00042-X

With mathematical population genetics in full flower, J.B.S. Haldane and H.T.J. Norton would publish two landmark papers, Norton (1928) and Haldane (1927) the latter influenced by the Norton article in manuscript form (Charlesworth, 2000). These articles developed basic models for how selection depends on "the Malthusian parameter," when there is age-structure. What sets apart Norton's 1928 paper is its expert handling of population genetics in relation to demography. Norton (1928) starts with a hypothetical diploid randomly mating population with a single locus and two alleles (A_iA_j). By varying the model parameters, Norton identified the key parameter that usually determines the eventual outcome of selection among alleles: the Malthusian parameter. The Malthusian parameter of a genotype A_iA_j, or the genotype's "r value," provides the eventual rate of population growth over time of a population made up exclusively of that genotype, as a function of $l_{ij}(x)$ survivorship and $m_{ij}(x)$ fecundity at age x, over a lifespan of length d. The exponential growth in population size as a function of time depends on r as defined by the following equation

$$n(t) = n(0)e^{rt} \tag{6.1}$$

The Malthusian parameter for genotype A_iA_j is the solution z from the next equation, one version of the Euler-Lotka equation (vid. Mueller, 2020):

$$\int_0^d e^{-zx} l_{ij}(x) m_{ij}(x) dx = 1 \tag{6.2}$$

These Malthusian parameters, Norton showed, determine the outcome of selection in the same manner as conventional fitness parameters from models of selection in populations that do not have age structure. For example, when genotypes bearing a particular allele always have lower Malthusian parameters than all other genotypes, mathematical analysis shows that particular allele would be eliminated from the population by natural selection.

By contrast, Norton showed that in the case of heterozygote superiority over homozygotes for Malthusian parameters, genetic variance is maintained by natural selection, with a polymorphic genetic equilibrium. This was an early anticipation of the possibility of selectively maintained genetic variation for life history, an idea not fully developed until Charlesworth (1980). The mathematical analyses of Norton and Haldane explained the importance of age for selection in populations with overlapping generations in a one-locus two-allele model with time-independent functions for life-history characters like age-specific survival.

Impact: 8

The work of Haldane and Norton demonstrated the importance of the Malthusian parameter in understanding how natural selection impacts fitness in an age-specific manner. It is likely that Fisher (see Chapter 7) was influenced by their work in using the Malthusian parameter to define fitness in his 1930 book.

Unfortunately for the evolutionary biology of aging, neither Norton nor Haldane applied their age-structured mathematical apparatus for natural selection to the problem of biological aging. Instead, for several decades their tools lay dormant, waiting to be taken up in service of the explicit mathematical analysis of the population genetics of aging.

References and further reading

Charlesworth, B. (1980). *Evolution in age-structured populations*. Cambridge, U.K: Cambridge University Press.

Charlesworth, B. (2000). Fisher, medawar, hamilton and the evolution of aging. *Genetics, 156*(3), 927.

Fisher, R. A. (1918). The correlation between relatives on the supposition of mendelian inheritance. *Transactions of the Royal Society of Edinburg, 53*, 399—433.

Fisher, R. A. (1930). *The genetical theory of natural selection*. Clarendon Press. https://doi.org/10.5962/bhl.title.27468

Haldane, J. B. S. (1927). A mathematical theory of natural and artificial selection, Part V: Selection and mutation. *Mathematical Proceedings of the Cambridge Philosophical Society, 23*(7), 838—844. https://doi.org/10.1017/S0305004100015644

Haldane, J. B. S. (1932). *The causes of evolution*. Macmillan.

Lotka, A. J. (1925). *Elements of physical biology*. Baltimore: Williams and Wilkins Company.

Mueller, L. D. (2020). *Conceptual breakthroughs in evolutionary ecology*. Elsevier.

Norton, H. T. J. (1928). Natural selection and mendelian variation. *Proceedings of the London Mathematical Society, s2—28*(1), 1—45. https://doi.org/10.1112/plms/s2-28.1.1

Weismann, A. (1889). *Essays upon heredity and kindred biological problems*. Oxford: Claredon Press.

Weismann, A., Poulton, E. B., Schönland, S., (Selmar), Shipley, A. E., & Arthur, E. (1891). *Essays upon heredity and kindred biological problems*. Oxford: Clarendon press. http://archive.org/details/essaysuponheredi18901weis

CHAPTER SEVEN

1930: First explanation of aging by age-specific patterns of selection

The standard paradigm

While Weismann was the first to link aging to patterns of natural selection, he did so without the tools and insights of theoretical population genetics. Norton (1928) first developed the theoretical population genetics framework that could be applied to the evolution of aging but did not in fact attempt such an application. Thus, before 1930, it would be fair to say that the evolutionary biology of aging was a field that hardly existed.

Unfortunately, the project of incorporating the mathematical tools of theoretical population genetics that Norton preferred with an evolutionary theory of aging would remain unaddressed until the 1970s. But the years from 1930 to 1970 saw the proposal of a variety of scientific concepts that were stepping stones toward the ultimate development of a satisfactory evolutionary analysis of aging. Those stepping stones will be our chief concern now.

The conceptual breakthrough

In his 1930 book *The genetical theory of natural selection*, Fisher was the first to link theoretical evolutionary genetics with the evolution of senescence. Unfortunately, his links were not developed by explicit mathematical derivation, which may explain their eventual abandonment.

In his book, Fisher proposed that declining survival with adult age could be due to the declining "reproductive value" of older individuals, where he defined reproductive value as the expected future contribution of an individual of a particular age to the future reproduction of a population, discounted according to the rate of population growth. This concept is somewhat analogous to "present value" in accounting.

Conceptual Breakthroughs in The Evolutionary Biology of Aging
ISBN: 978-0-12-821545-6
https://doi.org/10.1016/B978-0-12-821545-6.00043-1

Like Norton and Haldane, Fisher started with $l(x)$ defined as the "survivorship" or probability of survival to the age x, while he took $m(x)$ as the fecundity of a given individual at age x. As in earlier work, these age-dependent life-history characteristics in turn define a rate of population growth, the Malthusian parameter, which is the solution r in the standard Euler–Lotka equation

$$\int_{0}^{\infty} e^{-rx}\, l(x)m(x)\; dx = 1 \qquad (7.1)$$

With these parameters and definitions, reproductive value is obtained from the following integral Eq. (7.2),

$$v(x) = \int_{x}^{\infty} e^{-ry}\, l(y)m(y)\; dy\; e^{rx}/l(x) \qquad (7.2)$$

Fisher supplied a plot of reproductive value against age, which he compares with human age-specific mortality rates, as rendered in Fig. 7.1. He then wrote, "It is probably not without significance in this connexion that the death rate in Man takes a course generally inverse to the curve of reproductive value. The minimum of the death rate curve is at 12, certainly not far from the primitive maximum of the reproductive value; it rises more

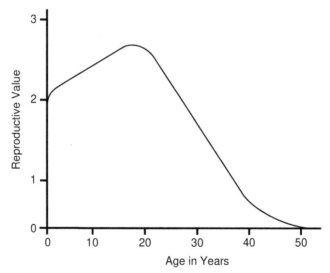

Figure 7.1 Illustration of Fisher's reproductive value, using demographic data of Australian women in a 1911 work. *Adapted from Fisher, R. A. (1958). The genetical theory of natural selection (2nd revised ed.).* New York: Dover Publications.

steeply for infants, and less steeply for the elderly than the curve of reproductive value falls, points which qualitatively we should anticipate, if the incidence of natural death had been to a large extent molded by the effects of differential survival" (Fisher, 1930, p. 28).

Thus Fisher laconically proposed that natural selection shapes both infant mortality and aging in accordance with a value metric of reproductive fitness. The basis of this intuition is presumably that the quantitative scale of age-specific reproductive values appears to determine how natural selection influences individual mortality rates.

Impact: 7

While his argument does suggest it, Fisher doesn't show *how* evolution by natural selection will shape patterns of age-specific mortality using explicit mathematical derivations. To model the connection between reproductive value and aging, Fisher plotted reproductive value versus mortality rates in a human cohort, in support of his somewhat vague proposal. While his hypothesis has quantitative elements, it is far from an explicit mathematical demonstration that his conclusions are in fact correct.

On the other hand, it is fair to say that the kind of argument that Fisher sketched was a marked improvement over the evolutionary reasoning supplied by Weismann. Furthermore, compared to the verbal rambling and unwarranted speculations of mainstream gerontology in the first half of the 20th Century, Fisher's appeal to a quantitatively measurable parameter like reproductive value as the key to the evolution of aging was a radical improvement. Implicitly, Fisher was building on Weismann's core concept that patterns of natural selection must be the chief determinant of patterns of aging. As such, his work sustained the idea of an evolutionary explanation of aging couched in terms of natural selection. He didn't himself supply a successful explanation, as we will demonstrate in further chapters, but he made that goal a prominent objective for evolutionary biology, as his 1930 book was a landmark achievement for the field of evolutionary theory as a whole.

References and further reading

Charlesworth, B. (2000). Fisher, Medawar, Hamilton and the evolution of aging. *Genetics, 156*(3), 927.
Fisher, R. A. (1930). *The genetical theory of natural selection*. Clarendon Press, Oxford.

Fisher, R. A. (1958). *The genetical theory of natural selection* (2nd revised ed.). New York: Dover Publications.

Haldane, J. B. S. (1941). *New paths in genetics* (1st ed.). George Allen and Unwin.

Lotka, A. J. (1925). *Elements of physical biology.* Baltimore: Williams and Wilkins Company.

Norton, H. T. J. (1928). Natural selection and Mendelian variation. *Proceedings of the London Mathematical Society, s2—28*(1), 1—45. https://doi.org/10.1112/plms/s2-28.1.1

Medawar, P. B. (1946). Old age and natural death. *Modern Quarterly, 1,* 30—56.

Medawar, P. B. (1952). *An unsolved problem of biology.* H.K. Lewis and Co.

Williams, G. C. (1957). Pleiotropy, natural selection, and the evolution of senescence. *Evolution, 11*(4), 398—411. https://doi.org/10.2307/2406060. JSTOR.

CHAPTER EIGHT

1941: First proposal of the general idea of declining force of natural selection

The standard paradigm

In most of the aging research community, aging continues to be viewed as a physiological process of breakdown that arises from unavoidable physical limitations inherent to life. This common paradigm does not take into account the potential for age-specificity of gene-action, age-specific effects that might be shaped by natural selection.

As we have seen, Fisher (1930) supplied the first intuitive argument that rested on some type of age-specificity. But he based his thinking on "reproductive value," a quantitative parameter that then had no known or demonstrated connection to the action of natural selection in evolving populations. It would take a reformulation of the problem of how natural selection might affect aging before genuine theoretical progress with the evolutionary biology of aging could be made.

The conceptual breakthrough

In a relatively obscure book, *New Paths in Genetics,* Haldane (1941) explored the limitations of natural selection's ability to purge pervasive afflictions like Huntington's Disease, or Huntington's Chorea as it was then known. Huntington's Disease is unusual among genetic diseases, in that it is caused by a single dominant allele that has uniformly devastating effects on the human central nervous system. These effects rarely commence before the age of 35, but they are progressive and ultimately fatal, with the course of the disease from onset to death taking 10 years or more. Despite its devastating effects, this is a relatively common genetic disease, even among present-day human populations.

Haldane proposed that the evolutionary genetic explanation for the prevalence of Huntington's Disease in human populations was its onset at late ages, ages at which most afflicted individuals in ancestral human societies

Conceptual Breakthroughs in The Evolutionary Biology of Aging
ISBN: 978-0-12-821545-6
https://doi.org/10.1016/B978-0-12-821545-6.00006-6

would have already reproduced. In effect, the concept is that the disease normally struck so late in adult life that it had a negligible net effect on reproduction.

By contrast, genetic disorders caused by single dominant mutations that kill children prior to reproduction are individually rare, as in the case of diseases like Hutchinson—Gilford's progeria. This type of progeria has been of great interest among the mainstream gerontological research community because children afflicted with it have the appearance of small elderly people. These children also have very early-onset of cardiovascular disease, baldness, and wrinkling. But they do not exhibit early onset of cancer or dementia, making them only partially an example of "accelerated aging" in humans.

But the important contrast between childhood progeria and Huntington's Disease, at least for evolutionary genetics, is the difference in their frequency. Natural selection stymies the spread of progeria but allows the spread of Huntington's Disease. This contrast illustrates the changing strength of the force of natural selection with age, in Haldane's view. The force of natural selection is strongest during the early ages devoted to development toward adulthood but then weakens progressively after the onset of reproduction.

This reasoning reveals the key role of the age-specificity of genetic effects in the action of natural selection on aging. By drawing attention to the importance of age specificity, Haldane implicitly bolsters the case for an evolutionary explanation of aging. Furthermore, Haldane introduces the concept of a waning force of natural selection acting on age-specific effects. He doesn't specify what such forces of natural selection are determined by formally, nor does he supply an explicit evolutionary genetic model for the dynamics of natural selection acting on such age-specific genetic effects. But the basic ideas that the evolutionary biology of aging would work on were proposed in his 1941 book.

Impact: 7

Haldane (1941) verbally attributed the relative prevalence of genetic disorders with age-specific effects to the declining forces of natural selection with age. But he did not mathematically define the forces of natural selection. Nor did he show explicitly how declines in the forces of natural selection would directly shape the prevalence of either genetic diseases or aging in general.

He did offer the idea that selection might favor genetic modifiers that postpone the onset of genetic disease. If this evolutionary mechanism were the central foundation of aging, one would expect a wide range of well-defined "genetic diseases" with late ages of onset, rather than the diffuse and ill-defined pathologies associated with aging. Clinical evidence however does not support this hypothesis of delayed onset of genetic diseases, as disorders that fit this scenario, like Huntington's Disease and idiopathic hemochromatosis, are not prevalent among the wide range of individual genetic diseases.

Haldane's discussion of the evolution of age-dependent genetic disorders did supply the key idea of the force of natural selection. But his ideas about the specific population genetic mechanisms of the evolution of aging did not prove to be valuable foundations for further work on the evolutionary theory of aging.

References and further reading

Albin, R. L. (1988). The pleiotropic gene theory of senescence: Supportive evidence from human genetic disease. *Ethology and Sociobiology, 9*(6), 371–382. https://doi.org/10.1016/0162-3095(88)90027-1

Fisher, R. A. (1930). *The genetical theory of natural selection.* Clarendon Press.

Haldane, J. B. S. (1941). *New paths in genetics* (1st ed.). George Allen and Unwin.

CHAPTER NINE

1946—57: Verbal hypotheses for the evolutionary genetics of aging

The standard paradigm

While the notion of populations ceding immortality to increase the chances of reproduction was considered by Weismann, he lacked the population-genetics tools to elucidate the evolution of aging. Haldane, Norton, and Fisher provided insights into the impact of age-specificity for evolution, but did not propose useful genetic mechanisms that might underpin the evolution of aging.

Thus, the evolutionary theory of aging was still woefully incomplete as of 1945. It lacked explicit mathematical development, especially compared to the relatively successful development of formal theory for the action of natural selection on genetic variation impinging on the evolution of fitness in populations lacking age-structure (Such populations have a single episode of reproduction, following a period of synchronized development. Natural populations which have these features include annual plants and univoltine insects).

The evolutionary theory of aging also lacked credible and specific hypotheses as to how genetic variation with age-specific effects could be shaped by evolution so as to establish biological aging. Bear in mind that conventional gerontology had no need of such hypotheses. In its conceptual framing, aging would occur simply through cumulative physiological processes of some type, perhaps like the breaking down of a car or a house over time, in the absence of sufficient maintenance.

It was the latter deficiency which was remedied in the period from 1946 to 1957.

The conceptual breakthrough

From 1946 to 1957, Peter B. Medawar and George C. Williams independently developed two distinct hypotheses concerning the genetics of the

Conceptual Breakthroughs in The Evolutionary Biology of Aging
ISBN: 978-0-12-821545-6
https://doi.org/10.1016/B978-0-12-821545-6.00024-8

evolution of aging, given the declining force of natural selection: (1) antagonistic pleiotropy between early life benefits and later-life costs for age-specific life-history characters and (2) accumulation of deleterious mutations with effects confined to late ages. Each concept, while not mutually exclusive, incorporated the idea of declining forces of natural selection in order to explain the evolution of aging.

The verbal hypothesis of "antagonistic pleiotropy," a term that neither Medawar nor Williams used, as an evolutionary genetic mechanism for the evolution of aging is based on opposing effects of genetic variants on early life versus late life. For example, Williams (1957) proposed that the "selection of a gene that confers an advantage at one age and a disadvantage at another will depend not only on the magnitudes of the effects themselves, but also the time of the effects." In this schema, a slight advantage at an earlier stage, closer to a population's first age of reproduction, could offset even the most deleterious of effects sufficiently later in life. This is due to the reduced impact on fitness at later ages, when natural selection is at its weakest, following the reasoning of Haldane in 1941 (Fig. 9.1).

Mutation accumulation, the other population-genetic mechanism proposed for the evolution of aging, is based on the hypothesis that aging might arise from the evolutionary accumulation of deleterious genetic effects that

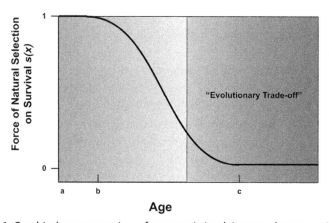

Figure 9.1 Graphical representation of antagonistic pleiotropy theory—aging manifests as a consequence of natural selection favoring genes that are beneficial in an organism's early life (A—B) despite deleterious effects on late life (C). Natural selection is strongest up to the reproductive phase of an organism's life and begins to decline with a progressive weakening of the force of natural selection. The visual gradients illustrated denote the trade-offs between benefits (*green* (light gray in print)) in early life versus the deleterious consequences (*red* (dark gray in print)), of the same favored genes, in late-life.

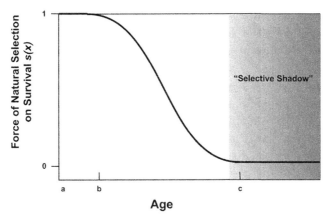

Figure 9.2 Graphical representation of mutation accumulation theory—aging manifests as a result of the declining force of natural selection, producing a "selective shadow" seen in *red* (gray in print). Mutations that would otherwise be filtered out by natural selection "build up" in late-life due to sufficiently weak forces of natural selection creating an evolutionary "blind spot." The deleterious mutations presenting in late life do not have benefits in early life, like those supposed by the antagonistic pleiotropy theory.

become phenotypically penetrant so late in adult life that they have no effect on Darwinian fitness (Medawar, 1952). In effect, such mutation accumulation is expected to arise when alleles have deleterious effects at ages so great that there is no remaining force of natural selection acting on them. Note that this hypothesis specifically requires an *absence* of pleiotropic effects between ages (Also note that this hypothesis has nothing to do with the hypothesis of "somatic mutation" causing aging, which we discuss in Chapter 13). Medawar proposed this idea entirely on his own in his 1946 and 1952 publications. Williams did not propose it in his 1957 article (Fig. 9.2).

Impact: 8

The publications of Medawar and Williams on the evolutionary genetic of aging were primarily verbal presentations of two alternative hypotheses, antagonistic pleiotropy, and mutation accumulation. There are no generally established principles from genetics which support or preclude these two hypotheses. Together, they created a foundation from which experiments would be designed and then conducted to falsify either proposition in question.

Before 1957, there was only circumstantial evidence for or against the patterns of pleiotropy that these two evolutionary genetic hypotheses presumed. As such they remained in play as mere possibilities. It was not until the work of Charlesworth (e.g., 1980) that these evolutionary genetic hypotheses were formulated in terms of explicit theoretical population genetics.

But once these hypotheses were formulated mathematically, they supplied the foundations for experimental testing and evaluation of whether evolutionary theory might account for the existence and patterns of aging. Thus these publications from Medawar and Williams, as lacking as they were in mathematical development, nonetheless proved seminal for the field.

References and further reading

Charlesworth, B. (1980). *Evolution in age-structured populations.* Cambridge, U.K: Cambridge University Press.

Medawar, P. B. (1946). Old age and natural death. *Modern Quarterly, 1,* 30—56.

Medawar, P. B. (1952). *An unsolved problem of biology.* H.K. Lewis and Co.

Medawar, P. B. (1955). The definition and measurement of senescence. In G. E. W. Wolstenholme, & M. P. Cameron (Eds.), *Ciba foundation symposium—General aspects (colloquia on ageing)* (Vol. 1, pp. 4—15). John Wiley and Sons, Ltd. https://doi.org/10.1002/9780470718926.ch2

Williams, G. C. (1957). Pleiotropy, natural selection, and the evolution of senescence. *Evolution, 11*(4), 398—411. https://doi.org/10.2307/2406060. JSTOR.

CHAPTER TEN

1953: Absence of a Lansing effect in inbred Drosophila

The standard paradigm

Despite numerous advances in the field, aging remains one of the few biological phenomena where Lamarckian notions of cumulative physiological effects have not only taken root but are commonly assumed.

An early example of the invocation of such cumulative physiological effects was the work of A. Lansing on rotifers (1947, 1948). Naturally parthenogenetic, *Philodina citrina* eggs were cultured by Lansing over multiple generations from older mothers, and thereby purportedly accumulated deleterious maternal effects. The eventual dying off of these cultures reared from older mothers was taken by Lansing as an example of cumulative physiological breakdown.

This "Lansing Effect" suggested that the reproduction of individuals exclusively at later ages would lead to accelerated aging. If such effects were widespread, then Lamarckian theories of aging would be more credible than the Darwinian theories that had been slowly developing in the trickle of publications from Weismann, Haldane, Medawar, and others.

Furthermore, the physiological reasoning underlying the inference of such Lansing Effects conformed well with the Aristotelian tradition of mainstream gerontology. After all, if such effects can accumulate across generations, it is all the more plausible that they have significant effects on the functional physiology of individual organisms during the course of their lives.

The Darwinian approach to the evolutionary biology of aging faced a mortal threat. Should reproducible evidence demonstrate the inheritance of physiological damage in the manner described by Lansing, the role of evolution in shaping aging would become less evident. Thus, experimental tests of Lansing's ideas became crucial to the survival of the nascent field.

Conceptual Breakthroughs in The Evolutionary Biology of Aging
ISBN: 978-0-12-821545-6
https://doi.org/10.1016/B978-0-12-821545-6.00027-3

The conceptual breakthrough

In an effort to test the findings of Lansing, Alex Comfort (1953) subjected an inbred line of *Drosophila subobscura* to eight generations of rearing from older females exclusively. No directional change in lifespan was found, and the populations did not crash as would be expected from a "Lansing effect." This absence of a reproducible and universal Lansing effect discredited the Lamarckian idea of cumulative physiological disruption spreading from generation to generation. Implicitly, it suggested that some other type of culture problem was the cause of the Lansing effect, not a Lamarckian effect. However, it is sociologically interesting that gerontologists were among the few who thought about physiology in Lamarckian rather than Darwinian terms, a pattern that continues to this day.

One explanation for why the Lansing populations died out could be that his rotifer cultures were not properly maintained, dying out due to poor environmental conditions. This interpretation is strengthened by the failure of subsequent studies (Meadow & Barrows, 1971) to reproduce the result, despite following the Lansing protocol. It should also be added that the rotifer cultures which Lansing reproduced from younger adults also died out. The natural explanation of Lansing's supposedly epochal findings is that he was not particularly proficient at maintaining rotifer cultures in his laboratory.

Impact: 5

Comfort's test of the Lansing effect was one of the first refutations of any anti-Darwinian hypothesis for aging. Although this paper only indirectly furthered the evolutionary explanation of aging, it falsified a potentially fatal challenge to the evolutionary theory of aging.

More generally, Comfort's work helped clear out some Lamarckian underbrush in the field of gerontology. It is not generally noticed by mainstream gerontologists that their core assumptions about aging as a strictly physiological process pose a significant challenge to many core ideas about life history, such as the physiological health of most offspring. But Comfort's work was an early and salutary attempt to clean up the field of gerontology, implicitly weaning it off speculations that had been long sustained over the preceding millennia since Aristotle.

This episode also highlights one of the most systemic problems in science: namely, poor replication and irreproducible findings slowing progress in a field. Lansing's work featured such poor experimental design that it proved difficult to untangle. Given the plethora of unreplicated and confusing studies performed in the field of gerontology, the task of reproducing or falsifying the field's myriad results is Sisyphean, where the effort to clear the intellectual tangles that these many studies produce becomes debilitating. Perhaps for that reason, mainstream gerontology has featured a seemingly endless procession of ideas that wax and wane without strong-inference support or falsification.

References and further reading

Comfort, A. (1953). Absence of a lansing effect in *Drosophila subobscura*. *Nature, 172*(4367), 83—84. https://doi.org/10.1038/172083a0

Lansing, A. I. (1947). A transmissible, cumulative, and reversible factor in aging. *Journal of Gerontology, 2*(3), 228—239. https://doi.org/10.1093/geronj/2.3.228

Lansing, A. I. (1948). Evidence for aging as a consequence of growth cessation. *Proceedings of the National Academy of Sciences of the United States of America, 34*(6), 304—310.

Meadow, N. D., & Barrows, C. H. (1971). Studies on aging in a bdelloid rotifer. II. The effects of various environmental conditions and maternal age on longevity and fecundity. *Journal of Gerontology, 26*(3), 302—309.

CHAPTER ELEVEN

1961: Presence of aging in a fish with continued adult growth

The standard paradigm

Bidder (1932) proposed that aging is a "cost" of terrestrial life. In his theory, the evolutionary transition to dwelling on land was responsible for the cessation of endless growth, as seen in some fish, because the constraints and challenges presented by gravity for terrestrial locomotion demanded mechanical efficiency. Aging was viewed by Bidder as "negative growth": "evolution [requires] some mechanism to stop natural growth" (Bidder, 1932) and aging is the regulatory mechanism required for this cessation. Bidder's ideas can thus be seen as natural extensions of Weismann's thinking from the previous century.

Bidder's postulations built on comparative biology, such as the unlimited growth potential of nonsenescent fissile organisms, in the absence of experimentation. Bidder speculatively extrapolated this pattern to explain the ostensible association between indefinite growth and indefinite longevity that he presumed to occur among some species of trees and fish. Although few fish species had at that time been studied thoroughly for longevity, there was existing evidence of fish growing larger postmaturity, with improved reproductive output at later ages. Likewise, trees were well-known to increase in reproductive capacity as they grew larger throughout their adult lives. Trees are in turn among some of the longest-lived organisms.

The conceptual breakthrough

Alex Comfort performed an aging experiment with lab-raised guppies of the species *Poecilia reticulata* in a 1961 study intended to test Bidder's hypothesis that aging does not exist when growth continues among adults. The aquarium treatment groups that Comfort used were the following: diet-restricted versus nonrestricted, large tank versus small tank, and breeding versus nonbreeding. Under Bidder's hypothesis, the guppies would be expected to remain ageless, which would be inferred by no decline in mortality while in the process of continued growth. The Bidder hypothesis

Conceptual Breakthroughs in The Evolutionary Biology of Aging
ISBN: 978-0-12-821545-6
https://doi.org/10.1016/B978-0-12-821545-6.00008-X
43

was effectively falsified when Comfort's data showed that fish with continued growth after reproduction exhibited increased mortality with increasing age. While this individual study does not necessarily speak for all known vertebrates, it falsified Bidder's proposal that growth cessation is the cause of aging. One possible explanation for the common inference for the "agelessness" of trees and large fish may be the difficulty of observing their aging in nature, not the complete absence of aging.

Impact: 4

The work performed by Comfort on aging in guppies provides yet another example of the direct experimental refutation of a conventional physiological hypothesis for aging. Each of these refutations provides further evidence against the merely physiological explanation of aging, but they do not directly further the evolutionary explanation of aging.

References and further reading

Bidder, G. P. (1932). Senescence. *British Medical Journal, 115*, 5831.

Comfort, A. (1961). The longevity and mortality of a fish (lebistes reticulatus peters) in captivity. *Gerontology, 5*(4), 209−222. https://doi.org/10.1159/000211060

Maynard Smith, J. (1962). Review lectures on senescence. I. The causes of ageing. *Proceedings of the Royal Society of London. Series B, Biological Sciences, 157*(966), 115−127 (JSTOR).

CHAPTER TWELVE

1966: Mathematical derivation of the forces of natural selection

The standard paradigm

The conventional physiological paradigm for aging assumes that natural selection favors unlimited survival and fertility. This position fails to consider the roles of age-specificity and the weakening forces of natural selection with adult age. A major impediment to general understanding and use of evolutionary theories for the cause of aging was the implicit, rather than explicit, mathematical definition for the weakening forces of natural selection prior to 1966.

Thus, despite some interesting verbal speculations from 1930 to 1957, evolutionary biologists had yet to formulate a convincing mathematical model for how the forces of natural selection weakened with adult age. While such a model would not be developed all at once in a single article, the transition from 1966 to 1980 can be regarded as the period when the evolutionary theory of aging finally came of age.

The conceptual breakthrough

Building on the theoretical work and ideas of Haldane, Norton, Fisher, Medawar, and Williams, W. D. Hamilton set out to derive the forces of natural selection in age-structured populations. In his seminal 1966 paper, Hamilton implicitly used partial derivatives of the Malthusian parameter to characterize the impact of natural selection on age-specific survival as $s(x)/T$, where T was a measure of generation length and $s(x)$ is given by (Eq. 12.1) as follows

$$s(x) = \sum_{y=x+1} e^{-ry} l(y) m(y) \tag{12.1}$$

The components of this equation consist of $l(y)$, the probability of surviving from birth to a given age y, and $m(y)$, the fecundity at said age. These survivorship and fecundity parameters can in turn be used to solve for the Malthusian parameter r in the Euler-Lotka equation (Eq. 12.2) [Note that

Conceptual Breakthroughs in The Evolutionary Biology of Aging
ISBN: 978-0-12-821545-6
https://doi.org/10.1016/B978-0-12-821545-6.00029-7

Hamilton (1966) used a discrete-time analysis, while earlier work was chiefly done using continuous-time analysis. As the duration of age-classes shrinks, these two models become quantitatively similar]. With the assumption of a homogenous population, the Malthusian parameter r provides the rate of population growth, and can be obtained from the following equation.

$$\sum_{y=1}^{\infty} e^{-rx}\, l(y)m(y) = 1 \qquad (12.2)$$

Hamilton (1966) uses several numerical examples to show that his measure of the force of natural selection acting on mortality rates makes quite different predictions from those of Fisher's (1930) reproductive value. For example, he constructs scenarios of life histories for which Fisher's hypothesis implies an absence of aging, while Hamilton's (1966) forces of natural selection predict its presence.

Derived from Hamilton's $s(x)$ function, Fig. 12.1 presents the age-specific force of natural selection acting on survival as a percentage of its full force. Natural selection is strongest at earlier ages (ages a to b), only declining after the beginning of reproduction (ages greater than b). This descent persists until the last age of reproduction (c) in a population's evolutionary history. From this last age of reproduction onward, there is a complete plateau at the zero bound with respect to natural selection's responsiveness to age-specific effects on survival probabilities.

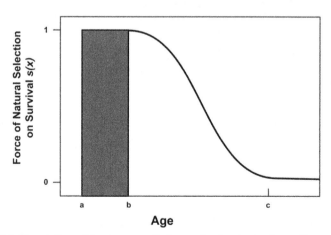

Figure 12.1 The scaling of the age-specific force of natural selection acting on survival falling with adult age, where a is the start of life, b is the start of reproduction, and c is the last age of reproduction in the evolutionary history of a population.

To mark the significance of this contribution, it is noteworthy that Eq. (12.1) provides the solution of some of the unanswered problems in prior chapters. For example, Hamilton's work established a mathematical foundation for Haldane's explanation for the high frequency of genetic diseases with late onset like Huntington's Disease, in terms of their selective unimportance. From Fig. 12.1, we can see that the effects of Huntington's Disease will impact survival probabilities so late in adult life that the disease has little effect on the Malthusian parameter, which usually defines fitness.

Likewise, this equation helps to explain the absence of aging noted in fissile organisms. The key evolutionary feature that permits the absence of aging in such organisms is that their forces of natural selection do not decline, because each act of symmetric fissile reproduction produces two offspring, *not* an adult and an immature offspring. In such organisms, there is no life-cycle after the onset of reproduction. Thus mathematically, there is no decline in its *s* function. Note, however, that this evolutionary effect requires the absence of physiological asymmetry between the products of fission. Vegetative reproduction which features strong physiological asymmetry, as in *Saccharomyces cerevisiae,* still permits the evolution of aging, because then there is an "adult phase," despite the absence of a "birth" process.

Hamilton (1966) derived a similar equation for the force of natural selection acting on age-specific fecundity, a result that again takes the form of a scaling function, $s'(x)$, divided by T, where T is the same measure of generation length,

$$s'(x) = e^{-rx}l(x) \tag{12.3}$$

and

$$\frac{\partial r}{\partial m(a)} = e^{-ra}\, l(a) \, / \, T \tag{12.4}$$

where

$$\sum_{x=1} xe^{-rx}\, l(x)m(x) = T$$

In the presented Formulae (12.3, 12.4), the force of natural selection will once again usually decline relative to age. In a stable or growing population, these functions will be negative due to the number of mothers aged $x+1$ being lower than the number of mothers aged x [Note that this pattern does not hold when a population is declining rapidly toward extinction.

In that case, offspring produced later will be disproportionately overrepresented. However, such populations are rarely observed, precisely because their numbers are falling rapidly toward extinction]. This will hold true for all ages past the first age of reproduction with respect to both fecundity and survival. Likewise, this also means that the selective pressures acting on changes to *l(x)* and *m(x)* in a population, such as from a mutation, will weaken across *x* after the start of reproduction. Both the pressure to keep a gene in a population as well as the pressure to purge a gene from a population will depend on the age of onset for its effects. Thus this partial mathematical analysis provides a formal setting for invoking Haldane's concept of declining forces of natural selection.

Impact: 9
The core evolutionary explanation of the biology of aging

Hamilton (1966) provided explicit mathematics underlying the fall in the forces of natural selection, an analysis missing from prior demographic work on senescence. His work serves to this day as an invaluable reference for the design of research that concerns the evolutionary biology of aging in the field of experimental evolution. By demonstrating the decline in the age-specific force of natural selection with age, when Darwinian fitness is tantamount to the Malthusian parameter, Hamilton provided a general explanation for the evolution of aging that is not dependent on a single physiological mechanism. Indeed, this type of theory allows the pervasive failure of adaptation across multiple functional characters.

What holds this breakthrough from a perfect impact score of 10 is that Hamilton (1966) derived the forces of natural selection from the assumption that the Malthusian parameter was equivalent to fitness in populations with age-structure, but he did *not* demonstrate that this presupposition was in fact the case. Nor did he discover the limitations on this presupposition. His mathematics thus only apply directly to the evolution of clonal populations with age structure.

The obvious next step in the development of the evolutionary theory of aging was to examine the extent to which Malthusian parameters could be used as accurate estimates of Darwinian fitness, because Hamilton's 1966 paper had made age-specific effects on those Malthusian parameters central to our understanding of the evolution of aging.

References and further reading

Charlesworth, B. (1980). *Evolution in age-structured populations*. Cambridge: Cambridge University Press.

Fisher, R. A. (1930). *The genetical theory of natural selection*. Clarendon Press.

Haldane, J. B. S. (1941). *New paths in genetics* (1st ed.). George Allen and Unwin.

Hamilton, W. D. (1966). The moulding of senescence by natural selection. *Journal of Theoretical Biology, 12*(1), 12—45. https://doi.org/10.1016/0022-5193(66)90184-6

Lotka, A. J. (1925). *Elements of physical biology*. Baltimore: Williams and Wilkins Company.

CHAPTER THIRTEEN

1960s: Falsification of the somatic mutation theory

The standard paradigm

The 1960s ushered in a paradigm shift in the field of gerontology, from a focus on the organism level to that of the cell. Cell-molecular theories of aging began to proliferate in the 1960s, inspired by in vitro studies of cell proliferation (e.g., Hayflick & Moorhead, 1961). That epochal publication established that normal vertebrate somatic cells, when cultured in glass vessels with nutritive serum, eventually slowed in their capacity to divide. Terminal, or "senescent," somatic cell cultures featured cells that continued to live, however. By various rhetorical sleights, Hayflick (e.g., 1965) connected this limited replicative capacity to the senescence of entire organisms. The fact that some arthropods, such as the entire taxon of insects, had little adult somatic cell division but considerable variation in lifespan was largely ignored.

One of the most interesting rhetorical devices used by cell-molecular biologists from the 1960s onward was to claim that they were only studying "cellular senescence" when they were challenged over the lack of direct association between cell proliferation and organ function. However, when other biologists aren't challenging them on that point, they have long elided any such distinction and referred to their research as the study of aging in general.

Having thus conveniently confused the relationship between cell proliferation in vitro and organismal aging in its entirety, mainstream biogerontology turned almost entirely to the question of which particular molecular mechanisms gave rise to limited cell proliferation. That, they held both explicitly and implicitly, was the central question for understanding biological aging.

One of the first such cell-molecular theories of aging was somatic mutation, which supposed that aging results from the cumulative dysfunction associated with progressive damage to the genomes of somatic cells over the lifespan of the organism (e.g., Curtis, 1966; Failla, 1960; Szilard, 1959).

Conceptual Breakthroughs in The Evolutionary Biology of Aging
ISBN: 978-0-12-821545-6
https://doi.org/10.1016/B978-0-12-821545-6.00046-7

The potential pervasiveness of such disruptions was thought to explain the myriad pathologies associated with aging, as physical damage to the DNA apparatus is plausibly going to disrupt organismal function globally. The primary challenge for this theory has been experimentally linking cellular mutation to reductions in the survival and reproduction of the entire organism.

One important version of the somatic mutation theory of aging was formulated by nuclear physicist Leo Szilard. In a 1959 article, Szilard provided a mathematical model that sought to calculate the influence of ionizing radiation on chromosomal damage and the resulting consequences for mean longevities. Szilard's hypothesized mutational damage takes the form of an "aging hit," which renders a chromosome and its genes inactive. Empirical studies of the impact of irradiation on vertebrate survivorship (vid. Lamb, 1977) produced a pattern quantitatively akin to that of aging. Instead of immediate increases in rates of death, like those produced by common stressors like starvation or desiccation, deaths due to irradiation quantitatively "shadow" or "echo" those of a normally aging population, but with some acceleration.

The conceptual breakthrough

Critical analysis of the impact of radiation and somatic mutation theory in general revealed that the similarities between radiation-induced death and normal aging were more superficial than mechanistically revealing. Two types of empirical data undermined this connection. 1. The detailed pathology of death due to radiation exposure featured accelerated incidences of cancer, but not accelerations of cardiovascular or other aging-associated diseases. 2. Patterns of aging in haploid and diploid animals did not match the theoretical expectations for somatic mutation.

While there is some evidence that somatic mutation theory can explain some cases for the onset of cancer, such as mutations to proto-oncogenes (Duesberg, 1983) responsible for regulating the proliferation of cells, the onset of such system-wide failures bears little resemblance to the patterns of aging beyond the superficial.

In studies performed by Maynard Smith (1958) and Strehler (1977, pp. 276–277), among others, the opposite result from that anticipated by somatic mutation theory was found: irradiated Drosophila lines lived longer than their control counterparts. In part, this may have been due to the benefits of sterilization associated with the effects of radiation (cf. Maynard Smith, 1958). In any case, these widely found experimental results demonstrated that somatic mutation is unlikely to be the sole and universal cause of aging.

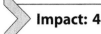

Impact: 4

It is worth noting that somatic mutation theory was not wholly refuted by research with radiation. But somatic mutation theory's claim to be a complete and sufficient explanation for patterns of aging was undermined. In principle, somatic mutation remains one of the best candidate theories cell-molecular biologists have for aging, though further studies (vid. Chapter 29) continue to reveal its limitations.

Despite decades of falsification of somatic mutation as the sole and sufficient cause of organismal aging, the status of the idea remains ambiguous among cell-molecular biologists. It is indubitable that somatic mutation occurs, and many studies have shown its occurrence among cells cultured in vitro (Albertini & Demars, 1973; Fulder, 1975; Finch, 1994). As such, it is thus the kind of cumulative physiological process of deterioration that Aristotle and all his successors have supposed to be nothing more or less than the entirety of aging.

Among the myriad cell-molecular mechanisms of aging that have been discussed, somatic mutation lives on in a kind of intellectual half-life. Other than cancer, for which somatic mutation is a demonstrable etiological factor, it has been hard to show that the many pathophysiologies of aging-associated disease, from cardiovascular disease to neurological degeneration, are caused by somatic mutation. But the field of mainstream biogerontology has shown little commitment to falsification or strong-inference experimentation generally. Thus somatic mutation often still crops up in lists of "mechanisms of aging" or "hallmarks of aging."

References and further reading

Albertini, R., & Demars, R. (1973). Somatic cell mutation detection and quantification of X-ray-induced mutation in cultured, diploid human fibroblasts. *Mutation Research/Fundamental and Molecular Mechanisms of Mutagenesis, 18*(2), 199—224. https://doi.org/10.1016/0027-5107(73)90037-7

Cole, L. C. (1954). The population consequences of life history phenomena. *The Quarterly Review of Biology, 29,* 103—137.

Curtis, H. J. (1966). The role of somatic mutations in aging. In P. L. Krohn (Ed.), *Topics in the biology of aging* (pp. 63—81). New York: John Wiley.

Duesberg, P. H. (1983). Retroviral transforming genes in normal cells? *Nature, 304*(5923), 219—226. https://doi.org/10.1038/304219a0

Failla, G. (1960). The ageing process and somatic mutations. In *The biology of ageing* (pp. 170—175). Washington, D.C: American Institute of Biological Sciences Symp.Publ. No.6.

Finch, C. E. (1994). *Longevity, senescence, and the genome.* University of Chicago Press.

Fulder, S. J. (1975). The measurement of spontaneous mutation in mammalian somatic cells. *Nucleus, 18,* 98—107.

Hayflick, L. (1965). The limited in vitro lifetime of human diploid cell strains. *Experimental Cell Research, 37*(3), 614–636. https://doi.org/10.1016/0014-4827(65)90211-9

Hayflick, L., & Moorhead, P. S. (1961). The serial cultivation of human diploid cell strains. *Experimental Cell Research, 25*(3), 585–621. https://doi.org/10.1016/0014-4827(61)90192-6

Lamb, M. J. (1977). *Biology of ageing*. Glasgow and London: Blackie.

Maynard Smith, J. (1958). The effects of temperature and of egg-laying on the longevity of Drosophila subobscura. *Journal of Experimental Biology, 35*(4), 832–842.

Strehler, B. L. (1977). *Time, cells and aging* (2nd ed.). New York: Academic Press.

Szilard, L. (1959). ON the nature of the aging process. *Proceedings of the National Academy of Sciences, 45*(1), 30–45. https://doi.org/10.1073/pnas.45.1.30

CHAPTER FOURTEEN

1960s: Falsification of the translation error catastrophe theory

The standard paradigm

Building off the momentum of the recent elucidation of the RNA-protein translation apparatus, Leslie Orgel (1963) proposed a new molecular explanation for aging that hypothesized a positive-feedback of errors in the synthesis of tRNAs and amino acyl synthetases. In a similar vein to somatic mutation theory, translation error catastrophe held the appeal of being a physiological aging mechanism with enough versatility to account for the great diversity of chronic age-related pathology. Although from an evolutionary perspective it would seem counterintuitive for RNA to be the driving force of aging, the autocatalytic nature of the process makes it one of the more plausible physiological theories of its era.

The core of translation error catastrophe is the proliferation of errors in the protein-synthesis apparatus, which in turn was expected to create a translation apparatus that was highly error-prone, with pervasive adverse physiological consequences. This theory's key proposition is delineated in Eq. (14.1), where the "error frequency" of generation n is represented by P_n, with discrete cell generations. The symbol k then represents the degree to which one cell generation contributes to this error frequency in generation $n+1$.

$$P_{n+1} = E + kP_n \tag{14.1}$$

Should the k value be > 1, the error frequency would increase to infinity. If it is less than 1, a stable state is possible and error catastrophe is avoided. While this model does not fully capture the complexity of the translation-apparatus, it conveys the consequences of sufficiently low translation fidelity as a root cause of the proliferation of error within a cell. This model naturally leads to the question as to how such a process would be allowed by evolution, unless the declining forces of natural selection are invoked. In any case, this ingenious theory was cogent enough to warrant empirical testing.

Conceptual Breakthroughs in The Evolutionary Biology of Aging
ISBN: 978-0-12-821545-6
https://doi.org/10.1016/B978-0-12-821545-6.00011-X

The conceptual breakthrough

Given the existence of potentially immortal fissile organisms (e.g., Bell, 1984), the translation-error theory of aging requires that nonaging species utilize a highly accurate apparatus for protein synthesis. But there is no evidence that this is in fact the case. Under the translation error catastrophe theory, the accumulation of nonfunctional proteins and worsening fidelity in amino synthesis with age would be two measurable characteristics associated with aging organisms. The absence of such cellular features would effectively refute the error catastrophe theory. Even supposing such features existed, it would not definitively prove this theory correct given the possibility of other biological explanations for said phenomena.

While an error catastrophe process was successfully generated in a mutant strain of *Neurospora* with a defective leucyl-t-RNA (Lewis & Holliday, 1970), it remains an isolated instantiation. Beyond this case, evidence for translation error-catastrophes could not be found in normal functioning cells (Holliday & Tarrant, 1972). Additional research with human fibroblast cultures did not reveal the predicted collapse in the fidelity of protein synthesis (Fulder & Tarrant, 1975). Loss of functional protein is relatively rare, but when it does occur it is most likely due to postsynthetic modification, not due to the breakdown of the protein synthesis machinery. An assay of translational fidelity monitored the ability of the *D. subobscura* protein-synthetic and transcriptional apparatus to distinguish between natural substrates and unnatural analogs (Shmookler Reis, 1976). There was no discernible shift in three discrimination ratios with age, and perturbations that markedly increased error rates to well above their normal levels in young or aged flies (elevated temperature or salinity) did not shorten lifespan, arguing against any role of translation error catastrophes in aging. While later studies would search for translation error catastrophes at higher levels of resolution (Gallant & Prothero, 1980), the observed frequency of translation errors was very low.

Impact: 5

The error-catastrophe theory stands as one of just a few cases where a cell-molecular theory of aging was investigated in enough detail, and with enough attention to falsification, that it is no longer seriously considered as a physiological explanation of aging. The contrast between the fates of the somatic mutation and error catastrophe cell-molecular theories is of historical, if not necessarily scientific, interest. Both somatic mutations and translation errors are virtually certain to occur, at whatever low frequencies. Yet the former process persists in lists of cell-molecular "aging mechanisms," while the latter does not.

Perhaps the key difference in their fates is that the translation error catastrophe theory made concrete predictions about the kind of cellular pathologies that it proposed produce aging. Specifically, a massive breakdown in translation fidelity would imply an exponentially increasing abundance of proteins with amino acid sequences that do not match genomic DNA. Such a collapse in cell physiology, however, is only observed when it is deliberately contrived through mutagenesis affecting translation fidelity (e.g., Lewis & Holliday, 1970). The translation error catastrophe theory is an example of a falsifiable, and thus useful, scientific theory that was indeed falsified. By contrast, the somatic mutation theory is an example of a theory that is much more difficult to pin down. Somatic mutations occur at low to moderate frequencies, but the proponents of the theory that they are a key "hallmark" of aging do not expose themselves to the scientific vulnerability of specifying exactly how somatic mutations are supposed to produce major aging diseases like those associated with cardiovascular systems.

The contrasting fates of the somatic mutation and translation error theories as "molecular mechanisms of aging" were instructive for the biologists who studied aging after the 1980s. In subsequent decades, the proponents of cell-molecular theories of aging would chiefly focus on a search for corroborative evidence in support of their theories. Attempts to falsify such theories would not be as common from the 1990s onward, and they would be still more rarely published, even when falsifying evidence was obtained. The cell-molecular theorists of aging had learned that experimental falsification of popular theories of their kind is a dangerous thing for their careers. With the failure of both the somatic mutation and error catastrophe theories of aging, strictly evolutionary theories of aging become of greater interest, at least among evolutionary biologists.

References and further reading

Bell, G. (1984). Evolutionary and nonevolutionary theories of senescence. *Am. Nat.*, *124*(4), 600–603.

Curtis, H. J. (1968). Biological mechanisms of delayed radiation damage in mammals. *Current Topics in Radiation Research, 3,* 139–174.

Fulder, S. J., & Tarrant, G. M. (1975). Possible changes in gene activity during the ageing of human fibroblasts. *Experimental Gerontology, 10*(3), 205–211. https://doi.org/10.1016/0531-5565(75)90033-9

Gallant, J. A., & Prothero, J. (1980). Testing models of error propagation. *Journal of Theoretical Biology, 83*(4), 561–578. https://doi.org/10.1016/0022-5193(80)90189-7

Hirsch, G. P. (1978). Somatic mutations and aging. In E. L. Schneider (Ed.), *Vol. 10. The genetics of aging* (pp. 91–134). New York and London: Plenum Press.

Holliday, R., & Tarrant, G. M. (1972). Altered enzymes in ageing human fibroblasts. *Nature, 238*(5358), 26–30. https://doi.org/10.1038/238026a0

Lewis, C. M., & Holliday, R. (1970). Mistranslation and ageing in Neurospora. *Nature, 228*(5274), 877–880. https://doi.org/10.1038/228877a0

Orgel, L. E. (1963). The maintenance of the accuracy of protein synthesis and its relevance to ageing. *Proceedings of the National Academy of Sciences of the United States of America, 49*(4), 517–521.

Shmookler Reis, R. J. (1976). Enzyme fidelity and metazoan aging. *Interdiscipline Topics and Gerontology, 10*, 11–23.

CHAPTER FIFTEEN

1968: Proposal of experimental designs to test evolutionary theories of aging

The standard paradigm

Before 1968, evolutionary biologists interested in connecting the declining forces of natural selection with aging believed their strongest evidence would come from interspecies comparative patterns, such as comparing the absence of aging among species reproduced by symmetrical fission. An example of such thinking was Comfort (1979), which described the absence of aging in an aquarium culture of fissile sea anemones. This was support for the explanation of the evolution of aging couched in terms of Hamilton's (1966) explicit formulation of the forces of natural selection, in that Hamilton's forces of natural selection do not decline with strictly symmetrical fissile reproduction.

The conceptual breakthrough

Edney and Gill (1968) opened the door to an experimental approach to testing evolutionary theories of aging. Their specific proposal was that the evolutionary theory of aging could be tested by changing the timing of reproduction in an outbred population for a number of generations, and then assaying the product of such experimental evolution for altered senescence. More specifically, Edney and Gill posited that delayed aging could be achieved by culturing lab populations from older individuals only, while accelerated aging could be achieved by culturing lab populations from younger individuals exclusively. In their opinion, the latter experimental design was more likely to yield useful experimental results within a shorter time frame, which is arithmetically reasonable, because more generations of experimental evolution are possible with shorter generation lengths.

Conceptual Breakthroughs in The Evolutionary Biology of Aging
ISBN: 978-0-12-821545-6
https://doi.org/10.1016/B978-0-12-821545-6.00032-7

Impact: 7

This publication was an early germ for subsequent research on the evolution of aging. While it does not provide direct empirical evidence, it is nonetheless directly tied to testing the evolutionary explanation of aging. Additionally, Edney and Gill (1968) outlined important concerns for implementing evolutionary studies on aging, such as reducing background levels of random mortality in experimental designs.

Note that their proposal pushed the evolutionary biology of aging away from historical evolutionary theories, like the speculations of Bidder (1932). Instead, their reasoning paved the way for strong-inference laboratory experiments, experiments that would place the evolutionary theory of aging at risk of falsification. It was the beginning of a strong "cultural" or methodological divergence from the practice of cell-molecular gerontologists, who would increasingly avoid experimental paradigms in which their ideas might be falsified.

References and further reading

Bidder, G. P. (1932). Senescence. *BMJ, 2*(3742), 583—585. https://doi.org/10.1136/bmj.2.3742.583

Comfort, A. (1958). Mortality and the nature of age processes (Alfred Watson memorial lecture). *Journal of the Institute of Actuaries, 84*(I—III).

Comfort, A. (1979). *The biology of senescence (Third)*. Churchill Livingstone.

Edney, E. B., & Gill, R. W. (1968). Evolution of senescence and specific longevity. *Nature, 220*(5164), 281—282. https://doi.org/10.1038/220281a0

Hamilton, W. D. (1966). The moulding of senescence by natural selection. *Journal of Theoretical Biology, 12*(1), 12—45. https://doi.org/10.1016/0022-5193(66)90184-6

CHAPTER SIXTEEN

1968: Accidental evolutionary postponement of aging

The standard paradigm

Some European biologists in the 1960s still believed that reproduction of individuals exclusively at later ages would lead to accelerated aging, the so-called the "Lansing Effect" from Chapter 10. Their continued belief was that Lamarckian effects were responsible for the dying off of Lansing's populations, despite the work of Comfort (1953).

The conceptual breakthrough

In order to test the Lamarckian-Lansing theory of aging, J. M. Wattiaux (1968a,b) performed two late-age reproduction experiments with Drosophila species, albeit with little replication and for a relatively small number of generations. In both cases, there was some increase in lifespan after multiple generations of late-age culture propagation, falsifying the Lansing-effect hypothesis. Of greater interest for the evolutionary biology of aging, these results suggested the feasibility of systematically using Hamilton's forces of natural selection to make populations with delayed aging evolve in lab cultures, as proposed by Edney and Gill (1968), although Wattiaux himself made no such connection (Fig. 16.1).

One of the experiments consisted of three different strains of *Drosophila pseudoobscura, AR/AR, CH/CH, CH/AR*. Flies designated as "T" were reared from younger parents (5—10 days of age), versus "B" flies from older parents (4 weeks). Flies were stored in bottles of ~200 flies with yeast changed every 4 days. In all longevity experiments, the older-parent culture regime led to higher adult survivorship among all crosses. He repeated this experiment across multiple species of Drosophila and found similar results, indicating similar patterns of evolution across species.

Conceptual Breakthroughs in The Evolutionary Biology of Aging
ISBN: 978-0-12-821545-6
https://doi.org/10.1016/B978-0-12-821545-6.00055-8

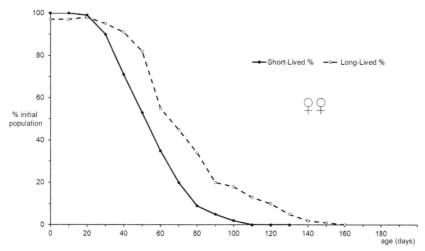

Figure 16.1 Female survival patterns of stocks of *D. subobscura*. The stock that was reproduced using young females is plotted as a solid line. The late reproduced stock is the dashed line, with greater longevities. *From Wattiaux, J. M. (1968a). Parental age effects in Drosophila pseudoobscura.* Experimental Gerontology, 3(1), 55–61. *https://doi. org/10.1016/0531-5565(68)90056-9.*

Impact: 6

Wattiaux viewed his findings as "controversial," perhaps because they didn't corroborate his non-Darwinian paradigm for aging that featured Lamarckian undertones. As an early empirical study demonstrating the power of manipulating the force of natural selection by changing the age of first reproduction, this study helped prepare the ground for later, entirely deliberate, experiments of similar design, albeit unintentionally. Future experiments would realize the potential latent in Wattiaux's work.

We note the historical irony that an attempt to corroborate a Lamarckian view of the evolution of aging resulted in yet another falsification of the view that the Lansing effect provided a general mechanism for aging. Furthermore, we note that it was another blow against the prevalent assumption that aging was underlain by some type(s) of cumulative physiological mechanism. Naturally, cell–molecular biologists took no note of this point, as they continued to drift toward weak inference experiments and attempts to corroborate their favored hypotheses.

References and further reading

Comfort, A. (1953). Absence of a Lansing effect in *Drosophila* subobscura. *Nature, 172*(4367), 83−84. https://doi.org/10.1038/172083a0

Comfort, A. (1958). Mortality and the nature of age processes (Alfred Watson memorial lecture). *Journal of the Institute of Actuaries, 84*(I−III).

Comfort, A. (1979). *The biology of senescence (Third)*. Churchill Livingstone.

Edney, E. B., & Gill, R. W. (1968). Evolution of senescence and specific longevity. *Nature, 220*(5164), 281−282. https://doi.org/10.1038/220281a0

Wattiaux, J. M. (1968a). Parental age effects in *Drosophila pseudoobscura*. *Experimental Gerontology, 3*(1), 55−61. https://doi.org/10.1016/0531-5565(68)90056-9

Wattiaux, J. M. (1968b). Cumulative parental age effects in *Drosophila subobscura*. *Evolution, 22*(2), 406−421. https://doi.org/10.1111/j.1558-5646.1968.tb05908.x

CHAPTER SEVENTEEN

1970: Experimental evolution of accelerated aging in *Tribolium*

The standard paradigm

Edney and Gill (1968) had proposed that sustained early cultures of outbred populations should lead to the evolution of accelerated aging. But they made no effort to perform such experiments themselves. Thus to this point in the development of the evolutionary biology of aging, the transition to powerful science had not yet been achieved. There was still no complete mathematical analysis of the evolution of aging. Nor were there strong inference experiments that had tested the burgeoning evolutionary theories of aging. Fortunately, the 1970s would mark the maturation of the evolutionary biology of aging. Perhaps surprisingly, an attempt at a strong inference experiment was one of the first steps toward the field's maturation.

The conceptual breakthrough

Sokal (1970) was the first to test Edney and Gill's proposal for an acceleration of senescence using experimental evolution with early culture reproduction. Four populations of *Tribolium*, flour beetles, were employed in the study. Two populations were raised for 40 generations using mass culture in which older individuals were permitted to reproduce. Two accelerated lines were subsequently derived from those mass culture lines, with the accelerated lines permitted to lay eggs for just 3 days at the start of adulthood before being killed and removed from each culture. The two beetle cultures maintained under the accelerated regime were then compared to their two respective ancestral stocks after 40 generations of the early culture selection regime.

Using paired comparisons of ancestral and accelerated populations, Sokal found that in each such comparison, one of the two genders from the accelerated populations had reduced longevity sufficient to reach statistical significance.

Conceptual Breakthroughs in The Evolutionary Biology of Aging
ISBN: 978-0-12-821545-6
https://doi.org/10.1016/B978-0-12-821545-6.00017-0

Impact: 6

These small-scale experiments suffered from a lack of replication and statistical power. Like Wattiaux's (1968a, b) research, previously described in Chapter 16, these early *Tribolium* experiments were not particularly well-designed. Both the limited size of the assayed cohorts and the small number of populations used, just four, made a strong result hard to achieve. Subsequent work in experimental evolution would show that the more replicate populations an experimenter employs, the greater the likelihood of reliable and strong inference (vid. Garland and Rose, 2009).

Nonetheless, these experiments provided early suggestive evidence that experimental evolution could definitively test the conjecture of Edney and Gill (1968), so long as artifacts like inbreeding were avoided. It should also be noted that this work fits with the similarly weak findings of Wattiaux (1968a, b), in suggesting that patterns of natural selection could shape the evolution of aging. This was in keeping with Weismann's original conjectures, as well as the general thrust of the developing mathematical approach to the evolution of aging. The stage was being set for the emergence of a field of significant scientific cogency.

References and further reading

Edney, E. B., & Gill, R. W. (1968). Evolution of senescence and specific longevity. *Nature*, *220*(5164), 281−282. https://doi.org/10.1038/220281a0

Garland, T., & Rose, M. R. (Eds.). (2009). *Experimental evolution*. Berkeley, California: University of California Press.

Mertz, D. B. (1975). Senescent decline in flour beetle strains selected for early adult fitness. *Physiological Zoology*, *48*(1), 1−23. https://doi.org/10.1086/physzool.48.1.30155634

Sokal, R. R. (1970). Senescence and genetic load: Evidence from Tribolium. *Science (New York, N.Y.)*, *167*(3926), 1733−1734. https://doi.org/10.1126/science.167.3926.1733

Wattiaux, J. M. (1968a). Parental age effects in *Drosophila pseudoobscura*. *Experimental Gerontology*, *3*(1), 55−61. https://doi.org/10.1016/0531-5565(68)90056-9

Wattiaux, J. M. (1968b). Cumulative parental age effects in *Drosophila subobscura*. *Evolution*, *22*(2), 406−421. https://doi.org/10.1111/j.1558-5646.1968.tb05908.x

CHAPTER EIGHTEEN

1970—74: Development of evolutionary genetics of age-structured populations

The standard paradigm

For organisms maintained in laboratory conditions, wholly discrete populations offer numerous experimental advantages. By creating systematically age-staged cohorts, like those of an American elementary school, the impact of a selection treatment is likely to be more uniform. Naturally, with discrete generations, it is straightforward to tabulate the number of generations that have elapsed since the start of an experiment. However, many species in the wild have overlapping generations, rather than discrete generations.

Populations with overlapping generations have complex age-structures, creating inherent difficulties for their mathematical modeling in demographic, ecological, and evolutionary theory. This mathematical challenge was taken up first by Haldane (1927) and Norton (1928), who provided an initial exploration of the mathematics required to represent the evolutionary genetics of populations with age-structure and overlapping generations. Their models employed a single-locus, two-allele system that featured selection. They also studied the attainment of Hardy-Weinberg equilibrium from specific starting conditions, with random mating.

Their findings demonstrated that, with relatively weak selection, populations with age-structure evolve in a similar fashion to populations with discrete generations, with fitness approximated by the Malthusian parameter introduced in Chapter 7. In the 1950s and 1960s, Motoo Kimura used the Malthusian parameter as an approximation for fitness in continuous-time models of evolving populations that incorporated stochastic effects. But those models were not developed in a way that was transparent enough for understanding the genetics of the evolution of aging. Notably, Hamilton (1966) too failed to use Norton's mathematics to explore the evolution of aging, instead proceeding directly to the use of the Malthusian parameter

Conceptual Breakthroughs in The Evolutionary Biology of Aging
ISBN: 978-0-12-821545-6
https://doi.org/10.1016/B978-0-12-821545-6.00048-0

for fitness. Hamilton (1966) did not provide explicit population-genetic models, only verbal scenarios for possible evolutionary-genetic patterns.

The conceptual breakthrough

A series of articles published by Brian Charlesworth in the 1970s reintroduced the work of Norton and Haldane, as well as providing greater clarity with respect to the mathematical machinery of evolutionary genetics with age-structure and thus overlapping generations. The breadth of Charlesworth's work requires us to organize our presentation under four subheadings: (1) conditions under which the Malthusian parameter fails to represent natural selection accurately; (2) conditions that allow genetic equilibrium; (3) invasion of new mutant genes in a population; and (4) derivation of dynamic equations for gene-frequency change.

1. Failings of the Malthusian Parameter

Charlesworth's 1970 treatment of the accuracy of using the Malthusian parameter as Darwinian fitness considered such scenarios as nonrandom mating, density-dependence, environmental change, and sex differentiation. Charlesworth found a number of cases in which the Malthusian parameter does not serve as an accurate surrogate for Darwinian fitness in age-structured populations. On the other hand, he found that the effect of an allele on the Malthusian parameter often provides a reasonable indication of how selection acts on that allele in a variety of cases, especially with weak selection and an absence of density-dependence. These latter results supported the cogency of the analysis of the evolution of aging by Hamilton (1966), namely the age-dependence of the effectiveness of natural selection on age-specific survival and fecundity.

2. Conditions for Evolutionary Equilibrium

Over a series of articles, Charlesworth (1972—74) further developed the mathematical machinery first pioneered by Norton in order to establish formal models that could accurately treat age-structured populations.

Above all, except for formally improbable cases, evolutionary genetic equilibrium requires age structure to converge on the stable population composition given by the eigenvector of the dominant eigenvalue, known as the Malthusian parameter, from the Euler-Lotka equation introduced in Chapter 7, where the terms of that equation are derived from the genotype-frequency weighted averages near genetic equilibrium (Charlesworth & Giesel, 1972).

In the simplest case, which is one locus with two alleles and no density-dependence, Charlesworth finds that there are relationships among appropriate fitness-measures which mirror the classic discrete-generation mathematics for such cases. If fitness-measures can be rank-ordered according to the number of alleles of a particular type, such that the more of one allele a genotype possesses the greater its fitness, there is no polymorphic evolutionary equilibrium. The allele that gives the greater fitness goes to fixation. It is only when the heterozygote has the greatest demographically weighted fitness that a polymorphic evolutionary equilibrium is maintained by natural selection.

3. Invasion of Mutant Gene

Charlesworth (1973) considered the case of gene frequency change for a gene with two variants, A_1 and A_2 having frequencies p and q, where q is the mutant allele that begins at an initially very low frequency, and the r_{ij} give the Malthusian parameter for the corresponding A_iA_j genotypes. A classic example of this would be a novel mutant gene possessing some beneficial effect on the Malthusian parameter entering a population at a low frequency.

Because q begins at an initial low frequency, when the invading allele affects the heterozygote Malthusian parameter the change in the frequency of q becomes

$$\frac{dq}{dt} \cong q(r_{12} - r_{11}) \tag{18.1}$$

The above equation shows that the progress of a rare variant through a population will be determined by the difference between the Malthusian parameter for the heterozygote and the Malthusian parameter of the homozygote. In order for a novel mutation to spread through a population, its effect on the Malthusian parameter of the heterozygote needs to be both beneficial and sufficiently strong.

4. Gene Frequency Change

Charlesworth (1974) derived equations that approximated the expected rates of gene frequency change and population growth in age-structured populations where density-dependence is not a factor.

The frequency of two variant alleles, for a locus with just those two alleles in a population, can be represented by p and q. For this specific case, there are three available genotypes, A_1A_1, A_1A_2, and A_2A_2 that will be referred to as 11, 12, and 22, respectively.

Let $B(t)$ be the number of new zygotes in a population at time t. Individuals in this populations have age x, ranging from the first day of reproduction, b, to the last day of reproduction, d. As is standard, $l(x)$ represents the probability or survival to age-x and $m(x)$ represents the fecundity of females at age-x. From these terms we find the following expression for $p(t)$.

$$p(t) = \int_b^d B(t-x)\, p(t-x)\{p(t-x)k_{11}(x) + q(t-x)\, k_{12}(x)\}\, dx / B(t)$$

$$(18.2)$$

where $k_{ij}(x) = l_{ij}(x)m_{ij}(x)$

The above equation gives an exact expression for the frequency of A_1 among zygotes conceived at time t. A similar equation also exists for $q(t)$ and A_2. As this equation represents the frequency of an allele at time t, further derivations were required for Charlesworth to go from the above equation to one involving the rate of change for allelic frequencies.

Unfortunately, an explicit relationship between $p(t)$ and $B(t)$ as a function of t cannot be obtained from the above equation. Due to this fact, approximations are necessary in order to produce useful theoretical results. After making key approximations allowable for the cases of constant population growth and weak selection, Charlesworth (1974) derives the following result

$$\frac{dp}{dt} = pq(r_{11\bullet} - r_{12\bullet})   (18.3)$$

where

$$r_{11\bullet} = pr_{11} + qr_{12}   (18.4)$$

and

$$r_{12\bullet} = pr_{12} + qr_{22}   (18.5)$$

Here r_{ij} measures the growth rate of a population whose members all share the same demographic characteristics as genotype ij. If we substitute these genotypic intrinsic rates of increase with Malthusian parameters, the above equation becomes analogous to the standard population dynamics with discrete generations. As Charlesworth states, "the present derivation, it is hoped, puts the use of the concept of the Malthusian parameter as the intrinsic rate of natural increase of a genotype on a sounder footing, as an extremely useful approximation to the case of weak selection" (Charlesworth, 1974).

Impact: 10

Without an explicit mathematical framework for the evolutionary genetics of populations with age-structure, progress toward understanding the evolution of aging would have remained haphazard, with biological intuition substituting for strong formal analysis. The latter, after all, was the practice of Haldane, Fisher, Medawar, and Williams in their publications on the evolution of aging. The work that Charlesworth did on the basic evolutionary genetic equations for age-structured populations made it possible for him, and others subsequently, to address the evolution of aging with appropriate stringency, rather than using the kind of intuitive appeals made by Fisher, Medawar, or Williams.

References and further reading

Charlesworth, B. (1970). Selection in populations with overlapping generations. I. The use of Malthusian parameters in population genetics. *Theoretical Population Biology, 1*(3), 352—370. https://doi.org/10.1016/0040-5809(70)90051-1

Charlesworth, B. (1973). Selection in populations with overlapping generations. V. Natural selection and life histories. *The American Naturalist, 107*(954), 303—311.

Charlesworth, B. (1974). Selection in populations with overlapping generations VI. Rates of change of gene frequency and population growth rate. *Theoretical Population Biology, 6*(1), 108—133. https://doi.org/10.1016/0040-5809(74)90034-3

Charlesworth, B., & Giesel, J. T. (1972). Selection in populations with overlapping generations. II. Relations between gene frequency and demographic variables. *The American Naturalist, 106*(949), 388—401.

Haldane, J. B. S. (1927). A mathematical theory of natural and artificial selection, Part V: Selection and mutation. *Mathematical Proceedings of the Cambridge Philosophical Society, 23*(7), 838—844. https://doi.org/10.1017/S0305004100015644

Hamilton, W. D. (1966). The moulding of senescence by natural selection. *Journal of Theoretical Biology, 12*(1), 12—45. https://doi.org/10.1016/0022-5193(66)90184-6

Kimura, M. (1958). On the change of population fitness by natural selection. *Heredity, 12*(2), 145—167. https://doi.org/10.1038/hdy.1958.21

Norton, H. T. J. (1928). Natural selection and mendelian variation. *Proceedings of the London Mathematical Society, s2—28*(1), 1—45. https://doi.org/10.1112/plms/s2-28.1.1

CHAPTER NINETEEN

1975: Application of Charlesworth's theory to the evolution of aging

The standard paradigm

Norton (1928) and Haldane (1927) mathematically examined the probability of a mutant gene's survival and its impact on a population's genetics. Hamilton (1966) had presumed that the long-term effectiveness of natural selection could be derived by taking partial derivatives of the Malthusian parameter. His analysis assumed that fitness is equivalent to the Malthusian parameter in age-structured populations. Hamilton (1966) did not, however, present explicit population-genetic models, only general-purpose verbal scenarios for possible evolutionary-genetic patterns. By focusing only on the Malthusian parameter, rather than developing explicit evolutionary genetic theory, Hamilton's work left a need for a new analysis that considered the impact of age-structure on the initial spread of a mutant with age-specific effects in a sexual Mendelian population, as a function of age.

The conceptual breakthrough

Charlesworth and Williamson (1975) mathematically analyzed the probability of survival of a mutant gene in an age-structured Mendelian system, as a function of age. They used a branching process model to analyze the fate of a mutant gene introduced into an age-structured population of large census size. Using a discrete-time model, they focused only on prereproductive and reproductive individuals, ignoring individuals who are no longer able to produce viable offspring. Here b denotes the first age of reproduction and d the last age of reproduction. P_x represents the probability of an individual surviving from age x to $x+1$, while l_x represents the probability of living to age x from birth. For the other important life history character, m_x represents half the number of offspring produced by an adult at age x, while M represents the mean total number of offspring across a generation. Given

Conceptual Breakthroughs in The Evolutionary Biology of Aging
ISBN: 978-0-12-821545-6
https://doi.org/10.1016/B978-0-12-821545-6.00021-2

that a new mutant will be introduced at age class 1 and that there remains a population of size N, a new mutant allele will first appear as a heterozygote. The probability of the genetic line of this individual surviving will be represented as u. All progeny produced by this new mutant are presumed be of the same generation, and thus the probability of its extinction becomes $q = 1-u$.

Charlesworth and Williamson begin with r, from the Euler-Lotka equation introduced in Chapter 6, a parameter which Hamilton referred to as the intrinsic rate of increase and has at other times in this book been referred to as the Malthusian parameter. A major goal of their paper was to apply age-structured evolutionary genetic theory to the problem of aging. They derived the following results for the case of u close to zero and the heterozygote's expected number of offspring, "M," near 1:

$$u \approx 2(M-1)/V \qquad (19.1)$$

$$r \approx (M-1)/T \qquad (19.2)$$

$$u \approx \frac{2rT}{V} \qquad (19.3)$$

To that end, the authors present u which represents the odds of survival for a new mutation in a population with stable demography. Here V represents the variance for the overall distribution of offspring across individuals in a generation and T is equal to the sum of the products of l_x and m_x values specific to the heterozygotes that carry the mutation, across all x ages. Before progressing, the authors also show how r can be approximated using the same parameters as u, and in certain circumstances can be used interchangeably. Additionally, these formulae show that changing M has more of an impact on u and r than manipulating the other variables.

The most important equations from the paper would be the partial derivatives for the relationship between the mutation survival probabilities, age-specific mortality, and age-specific fecundity, of which there are two expressions, one for prereproductive ages before an organism can have offspring, and one for reproductive ages, once organisms are capable of producing offspring.

(a) Prereproductive ages (x < b) For $u > 0$, we have

$$\frac{\partial u}{\partial \log p_x} \cong \left(l_b - \sum_{y=b}^{d} (l_y - l_{y+1}) H_y \right) \qquad (19.4)$$

When $u \approx 0$

$$\frac{\partial u}{\partial \log p_x} \cong M \tag{19.5}$$

H_x represents the total product of the generation function for all the heterozygous offspring produced by a mutant heterozygote across all days from day b to day x. It can be helpful to think of H as a representation of overall fecundity, rather than the age-specific fecundity of m. However, in this context H only refers to the total number of mutant offspring from a mutant parent.

In a similar fashion to Hamilton's own partial derivatives using r, the partial derivatives from this paper corroborate the idea that at prereproductive ages, the magnitude of u does not depend on the specific age-class a gene affects.

(b) Reproductive ages (d > x > b) For $u > 0$, we have

$$\frac{\partial u}{\partial \log p_x} \cong \left(l_{x+1} H_x - \sum_{y=x+1}^{d} \left(l_y - l_{y+1} \right) H_y \right) \tag{19.6}$$

When $u \approx 0$

$$\frac{\partial u}{\partial \log p_x} \cong \sum_{y=x+1}^{d} \left(l_y m_y \right) \tag{19.7}$$

Once organisms enter age-classes where reproduction becomes possible, the partial derivative for the above relationship involving u becomes a strictly decreasing function of x, because fewer positive terms appear in the sum defined on the right-hand side of the equation just given. This result shows explicitly that the force of natural selection acting on a mutant with age-specific effects on survival probability declines as the age x of such effects increases during the reproductive period, as Hamilton (1966) supposed it would.

Impact: 10

Charlesworth and Williamson (1975) showed that the fate of a newly introduced mutant allele in a Mendelian population depended on the age of its effects on age-specific life-history characters, during reproductive adulthood. This was the first satisfyingly complete analysis of the impact of the declining forces of natural selection on the genetic evolution of Mendelian populations. Numerical calculations in Charlesworth and Williamson (1975) show that the likelihood that a mutant allele with beneficial age-specific effects successfully invades an age-structured population is roughly parallel to Hamilton's Forces of Natural Selection, showing the same qualitative age-dependent pattern of decline.

Their result is not precisely equivalent to those obtained by Hamilton (1966), in their particular mathematical setting. However, they are considering a specific boundary case, rather than providing a complete mathematical analysis of allele frequency dynamics at the full range of frequencies, from near zero to near one, or fixation. Regardless, this was the first satisfyingly complete mathematical population-genetic analysis to support Haldane's original conjecture that the evolution of mutant alleles would depend on the age at which they have effects.

From this point onward, the evolutionary theory of aging developed like other branches of formal evolutionary theory, using similar tools to those applied to a variety of topics in the evolutionary genetics of natural populations. The evolutionary theory of aging is a fairly complicated body of applied mathematics, with a variety of special cases that require individual analysis. As with other fields of applied mathematics, recent decades have seen a transition away from explicit "algebraic analysis" toward numerical work exploring a wider variety of cases than can be analyzed without numerical specification. But in the 1970s, evolutionary theory had not yet made that transition. Thus, the work of Charlesworth and Williamson (1975) was of great importance in opening the door to an entirely formal approach to the development of the evolutionary theory of aging.

References and further reading

Charlesworth, B., & Williamson, J. A. (1975). The probability of survival of a mutant gene in an age-structured population and implications for the evolution of life-histories. *Genetics Research, 26*(1), 1—10. https://doi.org/10.1017/S0016672300015792

Haldane, J. B. S. (1927). A mathematical theory of natural and artificial selection, Part V: Selection and mutation. *Mathematical Proceedings of the Cambridge Philosophical Society, 23*(7), 838—844. https://doi.org/10.1017/S0305004100015644

Hamilton, W. D. (1966). The moulding of senescence by natural selection. *Journal of Theoretical Biology, 12*(1), 12—45. https://doi.org/10.1016/0022-5193(66)90184-6

Norton, H. T. J. (1928). Natural selection and mendelian variation. *Proceedings of the London Mathematical Society, s2—28*(1), 1—45. https://doi.org/10.1112/plms/s2-28.1.1

CHAPTER TWENTY

1980: Full development of evolutionary genetic theory for aging

The standard paradigm

The mathematics connecting age-structure to the evolution of aging in Mendelian populations had not been fully worked out before the 1970s. Two potential genetic mechanisms had been posited by 1960, where the evolution of aging is the result of: (1) *antagonistic pleiotropy*, in which alleles that have early benefits are favored by natural selection despite detrimental effects on later ages and (2) *mutation accumulation*, age specificity of gene action, where detrimental effects specific to later ages accumulate as a result of weakened selection at later ages, but are virtually neutral overall with respect to fitness. These mechanisms are not mutually exclusive, logically, and both could potentially influence the evolution of aging.

Both Peter Medawar and George Williams provided verbal descriptions of antagonistic pleiotropy, without using that particular term, and without providing fully developed mathematical theory, as described in Chapter 9. Mathematical formulation of the antagonistic pleiotropy hypothesis, together with its corollaries, had not yet been provided. The bare hypothesis was plausible, and its logic was simple enough. But its requirements and its consequences for experimentally testable attributes were not defined formally.

The alternative population genetic mechanism, mutation accumulation, was based on the absence of any pleiotropy between early and late ages, whether antagonistic or not. In its verbal formulation, evolutionary genetic theories of aging that are based on a presumed absence of pleiotropy between early and later life take three different forms.

First, the simplest of these nonpleiotropic formulations was *stasis of later adaptation*, initially proposed by Hamilton in (Hamilton, 1966). He supposed that the older individuals of a population are outperformed by the more selectively "up to date" younger individuals, and thus the older would evolve toward relative inferiority, but not absolute inviability. This theory

Conceptual Breakthroughs in The Evolutionary Biology of Aging
ISBN: 978-0-12-821545-6
https://doi.org/10.1016/B978-0-12-821545-6.00039-X

has been met with indifference. Perhaps the chief problem with it is that, even when older organisms are carefully cared for in zoos, arboretums, or homes, they evidently deteriorate to such an extent that they are difficult to keep alive, even with the ministrations of physicians or veterinarians.

Second, Haldane and Medawar had proposed that natural selection acting on conjectural modifier loci would postpone *the age of first onset of genetic diseases.* The problem with evolutionary genetic hypotheses of this kind is that they suffer from the low likelihood that such modifier alleles, which would provide benefits only for the small fraction of a population that has a particular genetic disease, would be selectively favored enough, relative to the rate of mutation at such modifier loci. Theories of this kind were promoted by Fisher and others in the 20th Century, but they are now considered implausible after further theoretical analysis. They will not be discussed further here.

The third and most cogent of these nonpleiotropic models is *mutation accumulation,* where this term refers to an evolutionary accumulation of deleterious effects later in adult life, due to recurrent mutations with strictly age-specific deleterious effects confined to later ages. In effect, this kind of theory implies that later adult life is an evolutionary genetic wastebasket that is not shaped by natural selection. This type of mutation accumulation is not to be confused with somatic mutations accumulating in cells later in the life of a single organism. Also note that if such late-acting alleles have positive or negative effects at early ages, when natural selection retains significant force, then the qualitative expectation is that such earlier genetic effects would predominate as determinants of the allele frequencies of later-acting alleles. Thus a strict absence of pleiotropy of any kind is required for this evolutionary genetic mechanism to work as proposed.

The conceptual breakthrough

Charlesworth's 1980 book *Evolution in Age-Structured Populations,* and subsequent 1994 edition, is responsible for developing rigorous population-genetic theory for age-structured populations, with a wide range of results concerning the effects of selection, conditions for genetic equilibrium, and so on. But our focus will be solely on those of its results that pertain to the evolutionary theory of aging. Charlesworth had previously developed formal theory for population genetics that featured both Mendelian inheritance and age-structure, as described here in Chapters 18 and 19. And his 1975 publication with Williamson provided an explicit treatment of a specific evolutionary genetic scenario by which aging patterns might be shaped. However, it is his 1980 book which provided solid and extensive foundations for the evolutionary genetics of aging.

Charlesworth (1980, pp. 206–217) provides one of the first satisfying mathematical treatments of antagonistic pleiotropy with the following example. The case that he analyzed involves an allele that impacts age-specific survival probabilities. The intensity of selection on age specific survival probability is expressed by $s_{ij}(x)/T$ for a given genotype A_iA_j at age x:

$$s_{ij}(x) = \sum_{y=x+1} e^{-r_{ij}y}\, l_{ij}(y) m_{ij}(y) \tag{20.1}$$

The numerator of the intensity of selection expression found in the partial derivative of Hamilton (1966)'s age-specific perturbations (Chapter 11) can be expressed similarly. In this scenario, the s(x) function denotes the age-specific intensity of natural selection on survival probability, which is scaled to the survival probability at the affected age. The T function is a measure generation length for the designated genotype A_iA_j.

$$T = \sum_{x=1} x e^{-r_{ij}x}\, l_{ij}(x) m_{ij}(x) \tag{20.2}$$

In order to approximate the impact of selection acting on this genotype with respect to fitness characters across all possible ages, Charlesworth developed a formula which substitutes a copy of each allele that makes up a diploid genotype. This is achieved by taking the differences between P_{ij} values when alleles are substituted and taking the weighted differences from the frequencies of other possible alleles (Falconer, 1981). With this setup, selection on the allele A_2 can be derived from its "average effect on gene substitution" at age x_k which is equal to the impact of changing one of these two alleles for the other, adjusted for the frequencies of the allele that is not undergoing substitution, as given by the following equation:

$$\propto_{kP} = p_1 \ln P_{12}(x_k) + p_2 \ln P_{22}(x_k) - p_1 \ln P_{11}(x_k) - p_2 \ln P_{12}(x_k) \tag{20.3}$$

Again, the p_i represents the gene frequencies of the A_i alleles and the P_{ij} are the age-specific survival probabilities associated with genotype A_iA_j. The kP subscript indicates that the effect is at age x_k and that the effect is on age-specific survival probability P. Using these α_{kp} variables, the equation for gene-frequency change can be written as the following, from Charlesworth (1980).

$$T_{11}\Delta P_2 \approx p_2(1-p_1)\left[\propto_1 p s_{11}(x_1) + \propto_2 p s_{11}(x_2)\right] \qquad (20.4)$$

This result shows that when the beneficial effects of an allelic substitution occur sufficiently early (at age x_1), relative to the deleterious effects (at age x_2), then selection is likely to favor such a substitution, with all prior assumptions being held consistent, because at such earlier ages the s scaling functions will be much greater in magnitude. Charlesworth outlines similar formulas for alleles affecting fecundity. Thus, alleles that present beneficial effects early, and only have their deleterious consequences sufficiently late, will be favored by natural selection. This mathematical formulation supplies explicit formal theory lacking in the verbal hypothesizing of Medawar and Williams.

Charlesworth (1980, pp. 140–142, 217, 218) goes on to develop explicit theory for the evolutionary genetic hypothesis of *mutation accumulation* arising from strict age-specificity of gene action, in the complete absence of pleiotropy. Here, his analysis focuses on the gene frequency equilibria at which there is a balance between recurrent deleterious mutation and selection against such deleterious alleles, when these alleles have effects strictly confined to survival probability at a specific age. In the proposed scenario, a mutant allele is considered that is not fully recessive, with onset of effects taking place at age x. The magnitude of the frequency of heterozygotes carrying the mutant allele (most of the individuals affected by the allele will be heterozygotes) is given by the following:

$$-2u/\left\{[lnP_{12}(x) - lnP_{11}(x)]s_{11}(x)\right\} \qquad (20.5)$$

Here, u is the rate of recurrent mutation, the P functions define the age-specific survival probabilities for the genotypes as before, and $s_{11}(x)$ is the numerator of the intensity of selection expression, already published in Hamilton (1966)'s treatment of the selective effects of age-specific perturbations (Chapter 11).

Impact: 10

This work developed full mathematical treatments of the alternative genetic hypotheses for the evolution of aging: (1) evolutionary accumulation of late-acting deleterious mutations; and (2) antagonistic pleiotropy between early and late genetic effects on life-history, leading to active selection for alleles that foster aging. With the mathematical foundation established by this book from Charlesworth, the design and interpretation of empirical tests that could corroborate or falsify either proposed mechanism for the evolutionary explanation of aging became relatively straightforward.

After 40 years of verbal speculation and incomplete mathematical analysis, Charlesworth made the alternative evolutionary genetic hypotheses for the evolution of aging mathematically explicit. This achievement did not, in itself, indicate whether the two obvious theoretical alternatives were correct. But it did direct Charlesworth and others toward the design and execution of strong-inference experiments that might refute or corroborate these bare possibilities for the evolution of aging.

Viewed still more broadly, we believe that it is fair to say that Charlesworth's book marked 1980 as the point in scientific history when the evolutionary theory of aging became a major contender for the status of a central unifying theory for understanding, explaining, and ultimately manipulating patterns of aging. It still remained to perform strong-inference experiments on the ideas of that theory. For any mathematical theory, however intuitively attractive, may prove to be erroneous in the cold light shone by powerful experiments.

References and further reading

Charlesworth, B. (1980). *Evolution in age-structured populations*. Cambridge: Cambridge University Press.

Charlesworth, B. (1994). *Evolution in age-structured populations* (2nd ed.). Cambridge University Press.

Falconer, D. S. (1981). *Introduction to quantitative genetics* (2nd ed.). London: Longman.

Hamilton, W. D. (1966). The moulding of senescence by natural selection. *Journal of Theoretical Biology, 12*(1), 12−45. https://doi.org/10.1016/0022-5193(66)90184-6

CHAPTER TWENTY ONE

1980−81: Quantitative genetic tests of hypotheses for the evolution of aging

The standard paradigm

The alternative hypotheses of mutation accumulation and antagonistic pleiotropy for the evolution of aging had been proposed in the 1940s and 1950s. Given that the field of quantitative genetics generally finds genetic variation in nearly all phenotypic characters, especially in outbred populations (Falconer, 1981; Mousseau & Roff, 1987; Roff & Mousseau, 1987), the intuition was that there would be such variation for aging as well. Work performed by Maynard Smith (1959) provided early evidence of quantitative genetic variation for longevity in Drosophila. This raised the possibility of testing the two leading alternative evolutionary genetic mechanisms for the evolution of aging, antagonistic pleiotropy and mutation accumulation, using standing quantitative genetic variation in the genus.

Previous chapters thus far have detailed the verbal arguments and theoretical mathematics concerning age-specific life history characteristics. But there is a conceivable hypothesis that is inimical to the evolutionary theory for aging developed from Haldane to Charlesworth. That hypothesis is an *absence of selectable age-specific genetic variation,* contrary to the suppositions of both the mutation accumulation and the antagonistic pleiotropy mechanisms for the evolution of aging. Mutation accumulation evidently leads to the maintenance of genetic variation due to selection-mutation balance, as we will discuss further below. Less noticed, initially, is that antagonistic pleiotropy leads to the selective maintenance of genetic variation, in cases where heterozygote fitness is greater than the fitness of homozygotes, which we will also discuss. Thus the alternative evolutionary genetic mechanisms for the evolution of aging make a simple core prediction that is eminently testable: abundant genetic variation for age-specific life-history characters, like survival, segregating in natural populations that have not been inbred. Put another way, if genetic variants with age-specific effects *never arise in*

Conceptual Breakthroughs in The Evolutionary Biology of Aging
ISBN: 978-0-12-821545-6
https://doi.org/10.1016/B978-0-12-821545-6.00012-1

natural populations, then it is not possible for any of the evolutionary genetic mechanisms for the evolution of aging to work.

In an evolutionary scenario with the complete absence of age-specific genetic effects, alleles are either globally beneficial or globally deleterious. Despite the abstract possibility of natural selection discriminating between the evolution of early life-history characters versus later life-history characters, it would not do so because these characters would be strictly positively correlated with each other genetically. The implication being that, with the correlations among all components of fitness being positive, the evolution of aging would be irrevocably tethered to fitness itself, as proposed by Giesel (Giesel, 1979; Giesel & Zettler, 1980). Experimental evidence supporting such an argument would need to demonstrate strong positive genetic covariation across all fitness components. The two proposed evolutionary genetic mechanisms for aging in terms of Hamilton's forces of natural selection, by contrast, require smaller positive, perhaps approximately zero or even negative, genetic correlations among life-history characters across ages.

The conceptual breakthrough

In 1980, it was thought that the mutation accumulation hypothesis implied that the additive genetic variance for age-specific survival or fecundity should increase massively with adult age. The aim of Rose and Charlesworth (1980, 1981a) was to use a quantitative-genetic sib analysis to determine estimates of additive genetic variance for daily fecundity in *Drosophila melanogaster* from early to later adulthood. If mutation accumulation were a strong determinant of the evolution of aging, they expected that such additive genetic variances should increase dramatically with adult age. But Rose and Charlesworth (1980, 1981a) found no such effect; rather, genetic variances for age-specific fecundity were stable with adult age (Fig. 21.1).

On the other hand, antagonistic pleiotropy implies that there should be negative genetic correlations between early fitness-characters and longevity. Such a negative genetic correlation was found by Rose and Charlesworth (1980, 1981a) in a quantitative-genetic "sib analysis." The conclusion drawn by them was that aging in these Drosophila was affected chiefly by alleles that featured late-acting deleterious effects together with early beneficial effects, alleles that were thus favored strongly by natural selection (Table 21.1).

This marked the first time evolutionary genetic speculations and mathematical models for aging received strong inference tests that could

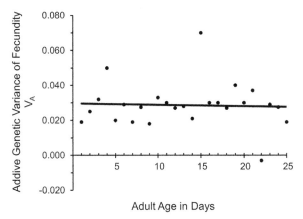

Figure 21.1 Estimates of the additive genetic variance V_A from daily fecundity days 1—25, relative to the mean. *Rose, M.R., & Charlesworth, B. (1981a). Genetics of life history in Drosophila melanogaster. I. Sib analysis of adult females.* Genetics, 97(1), 173—186. https://doi.org/10.1093/genetics/97.1.173.

Table 21.1 Estimated additive genetic correlations for adult female life-history characters in *D. melanogaster*.

	Fecundity days 1—5	Fecundity days 6—10	Fecundity days 11—15	Longevity
Fecundity days 1—5	—	−0.16	−0.48	−1.43
Fecundity days 6—10	−0.16	—	0.51	0.30
Fecundity days 11—15	−0.48	0.51	—	−0.71
Longevity	−1.43	0.3	−0.71	—

Data from Rose, M. R., & Charlesworth, B. (1981a). Genetics of life history in *Drosophila melanogaster*. I. Sib analysis of adult females. *Genetics, 97*(1), 173—186.

distinguish between the two proposed genetic hypotheses for the evolution of aging. In addition, these quantitative genetic results from Rose and Charlesworth (1980, 1981a) were strictly opposed to those from the Giesel laboratory. This was an early example of contrasting results from laboratories employing different experimental designs to test evolutionary genetic hypotheses. As we will discuss below, this kind of contrast would prove to hinge on the effects of experimental design choices on the results obtained from them. In particular, it is difficult, rather than easy, to conduct strong-inference experiments pertaining to the evolutionary genetics of aging when using quantitative genetic techniques of data analysis.

Impact: 8

Although the statistical power of this pioneering study from Rose and Charlesworth (1980, 1981a) was relatively modest, the key sib analysis was replicated seven times, with more than 1000 female fruit flies characterized. This study led to subsequent work on the quantitative genetics of aging (e.g., Hughes & Charlesworth, 1994; Promislow et al., 1996; Tatar et al., 1996). The overall course of this line of work can be summarized broadly here. It is difficult to connect patterns of genetic variance as a function of age with specific evolutionary genetic hypotheses. It is somewhat easier to derive useful conclusions from patterns of genetic covariation, though that too is fraught with artifacts and difficulties of interpretation.

References and further reading

Edney, E. B., & Gill, R. W. (1968). Evolution of senescence and specific longevity. *Nature*, *220*(5164), 281−282. https://doi.org/10.1038/220281a0

Falconer, D. S. (1981). *Introduction to quantitative genetics* (2nd ed.). London: Longman.

Giesel, J. T. (1979). Genetic co-variation of survivorship and other fitness indices in *Drosophila melanogaster*. *Experimental Gerontology*, *14*(6), 323−328. https://doi.org/10.1016/0531-5565(79)90044-5

Giesel, J. T., & Zettler, E. E. (1980). Genetic correlations of life historical parameters and certain fitness indices in *Drosophila melanogaster*. R m, rs, diet breadth. *Oecologia*, *47*(3), 299−302. https://doi.org/10.1007/BF00398520

Hughes, K. A., & Charlesworth, B. (1994). A genetic analysis of senescence in Drosophila. *Nature*, *367*(6458), 64−66. https://doi.org/10.1038/367064a0

Maynard Smith, J. (1959). Sex-limited inheritance of longevity in *Drosophila subobscura*. *Journal of Genetics*, *56*(2), 227. https://doi.org/10.1007/BF02984746

Mousseau, T. A., & Roff, D. A. (1987). Natural selection and the heritability of fitness components. *Heredity*, *59*(2), 181−197. https://doi.org/10.1038/hdy.1987.113

Promislow, D. E. L., Tatar, M., Khazaeli, A. A., & Curtsinger, J. W. (1996). Age-specific patterns of genetic variance in *Drosophila melanogaster*. I. Mortality. *Genetics*, *143*(2), 839−848. https://doi.org/10.1093/genetics/143.2.839

Roff, D. A., & Mousseau, T. A. (1987). Quantitative genetics and fitness: Lessons from Drosophila. *Heredity*, *58*(1), 103−118. https://doi.org/10.1038/hdy.1987.15

Rose, M., & Charlesworth, B. (1980). A test of evolutionary theories of senescence. *Nature*, *287*(5778), 141−142. https://doi.org/10.1038/287141a0

Rose, M. R., & Charlesworth, B. (1981a). Genetics of life history in *Drosophila melanogaster*. I. Sib analysis of adult females. *Genetics*, *97*(1), 173−186.

Rose, M. R., & Charlesworth, B. (1981b). Genetics of life history in *Drosophila melanogaster*. II. Exploratory selection experiments. *Genetics*, *97*(1), 187−196.

Tatar, M., Promislow, D. E. L., Khazaeli, A. A., & Curtsinger, J. W. (1996). Age-specific patterns of genetic variance in *Drosophila melanogaster*. II. Fecundity and its genetic covariance with age-specific mortality. *Genetics*, *143*(2), 849−858. https://doi.org/10.1093/genetics/143.2.849

1980–84: Mitigation of aging by postponing the decline in forces of natural selection

The standard paradigm

In order to achieve robust science, there must be a fruitful cycle between the development of formal theory and strong-inference tests of that theory. In a few classic cases from the history of science, such as the heliocentric theory of the solar system and Darwin's theory of evolutionary trees, formal theory may be fairly simple. Indeed, in the case of both the Copernican solar system and Darwin's evolutionary trees, the core theory may be embodied by quite simple graphs, without formal theory. In such cases, very simple but strong empirical findings may usefully test these straightforward theories.

But for most natural science, robust theoretical development requires the use of mathematics. Galileo and Newton contributed mathematical machinery that explained both terrestrial locomotion and astrophysical orbits, with subsequent research on locomotion and astrophysics relying still more on mathematical theory. Similarly, the development of Darwinian theory after 1900 proceeded with the proposal and analysis of important types of formal theory in the hands of R.A. Fisher, J.B.S. Haldane, and many others. The evolutionary theory of aging is no exception to this pattern, with more recent mathematics for the evolution of aging achieving high levels of difficulty (e.g., Mueller et al., 2011).

However, we believe that it is fair to say that, with the publication of Charlesworth's 1980 book, formal theory for the evolution of aging had reached a reasonable level of maturity, as argued in Chapter 20. While Charlesworth (e.g., 1980; personal communication) hoped that quantitative genetic analysis of patterns of genetic variance and covariance would be immediately useful in the empirical evaluation of his theoretical work, that hope has still not been fully realized, as is discussed throughout the present volume.

Conceptual Breakthroughs in The Evolutionary Biology of Aging
ISBN: 978-0-12-821545-6
https://doi.org/10.1016/B978-0-12-821545-6.00047-9

Instead, the strongest connections between theory and experiment for the evolution of aging would be forged within the field of "experimental evolution," a field that chiefly burgeoned after 1980 (vid. Rauser et al., 2009). While the inadvertent selection experiments of Wattiaux (1968a, b) suggested that reproducing outbred populations soley at later ages might lead to the evolutionarily postponement of aging, that was never his intent. The question remained: could experimental evolution be predictably tuned to produce delayed or slowed aging simply by postponing the first age of reproduction in an outbred population, as the mathematical work of Hamilton (1966) and Charlesworth (1980) suggested, and as Edney and Gill (1968) explicitly predicted?

In addition, if mechanisms of antagonistic pleiotropy for the evolution of aging were correct, then the postponement of aging by experimental evolution should be reproducibly associated with a reduction in some early life functional characters, such as viability, developmental speed, early female fecundity, or early male virility. By contrast, if the evolutionary genetics of aging depended chiefly on mutation accumulation, then the postponement of aging by experimental evolution should lack any such cost during early life. Thus experimental evolution supplied not only the opportunity to test Hamilton's equations; it also supplied the opportunity to test the alternative genetic mechanisms for aging proposed by Medawar (1946, 1952) and Williams (1957).

The conceptual breakthrough

Rose and Charlesworth (1980, 1981b) performed an unreplicated experiment specifically intended to evolve longer-lived Drosophila by increasing the first age of reproduction. They derived test lines by collecting eggs produced by females that were able to survive to day 21 from pupal emergence. That is, they moved the age of first reproduction from 14 days to 35 days from oviposition, while the ancestral, and control, "IV" population was maintained with reproduction at 14 days. After a dozen generations of delayed age at first reproduction, the two populations were compared with parallel rearing. This experiment was deliberately intended to test *both* the equations for the evolution of aging provided by Hamilton (1966) *as well as* the evolutionary genetic mechanisms proposed by Medawar (1946, 1952) and Williams (1957).

The offspring derived from the delayed-breeding populations lived approximately 10% longer than flies derived from their ancestral population

that continued to be bred at early ages. Of interest for the antagonistic pleiotropy hypothesis, these longer-lived fruit flies also showed reduced early
fecundity. Unfortunately, this work was unreplicated at the population
level, much like the work of Wattiaux (1968a, b).

Experiments with proper replication would soon follow. Luckinbill et al.
(1984) and Rose (1984) recapitulated the delayed-first-reproduction experimental design with replicated controls and replicated selection lines, finding
qualitatively similar results to those of Rose and Charlesworth (1980,
1981b). In the experiment led by Luckinbill, twofold replication of the
delayed-breeding population was used (Fig. 22.1). Rose (1984) followed a
similar procedure but used five replicate populations for each of the early
reproduction ("B") and late-reproduction ("O") selection protocols, though
only three replicates were initially analyzed (Rose, 1984).

Like the Luckinbill design, in the Rose (1984) study the age for first-
reproduction by females was progressively delayed, until the five O populations were routinely cultured using females that had lived about 70 days,
while the 5 B controls were maintained at the ancestral age of reproduction

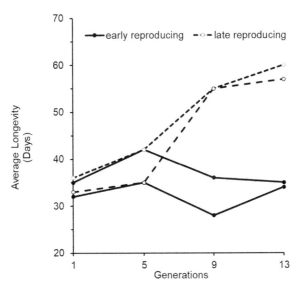

Figure 22.1 The sequential patterns of mean longevity change in stocks of
D. melanogaster maintained using older females, compared with controls. Mean
longevity consistently increases in the late-reproduced lines. *From Luckinbill, L. S.,
Arking, R., Clare, M. J., Cirocco, W. C., and Buck, S. A. (1984). Selection for delayed senescence in* Drosophila melanogaster. *Evolution, 38(5), 996—1003. JSTOR. https://doi.org/10.
2307/2408433*

at 14 days. The three O and three B populations compared by Rose (1984) supplied a statistically sounder corroboration of the original findings of Rose and Charlesworth (1980, 1981b): mean adult lifespan was increased, and early female fecundity was reduced in the O populations compared to the B populations (Figs. 22.2 and 22.3).

Impact: 9

Taken together, these four publications firmly established the principle that experimental evolution could be reliably used to postpone aging, in accordance with the expectations of the evolutionary theory of aging. Building off the predictions of Edney and Gill (1968), each paper discussed here provides empirical evidence for the theoretical framework of selection acting powerfully on survival early in life, then subsequently falling with the initiation of reproduction.

Note that these three experiments together provided a strong-inference test of the core results from Hamilton (1966) and Charlesworth (e.g., 1980): the evolution of adult survival depended critically on the timing of the start of actual reproduction by an outbred population with standing genetic

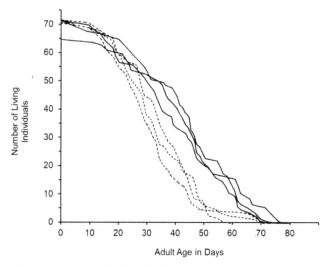

Figure 22.2 Survival patterns for females from stocks of *D. melanogaster* maintained using females of different ages. Stocks reproduced using older females are plotted as solid lines. Stocks reproduced using younger females are plotted as dashed lines. *From Rose, M. R. (1984). Laboratory evolution of postponed senescence in* Drosophila melanogaster. Evolution, 38(5), 1004–1010. JSTOR. https://doi.org/10.2307/2408434

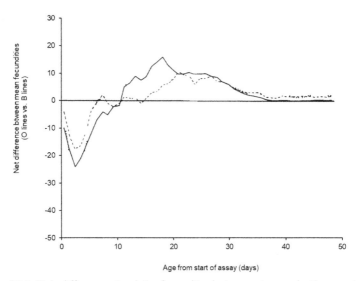

Figure 22.3 Net difference in daily fecundity between two selection regimes of *D. melanogaster*, the late reproducing O line versus the early reproducing B lines. Each line variant (e.g., solid or dashed) represents a different pairing of a replicate of O versus B. At early ages the O lines demonstrate relatively depressed early fecundity, but at later ages the O lines have notably enhanced late fecundity. *Data from Rose, M. R. (1984). Laboratory evolution of postponed senescence in* Drosophila melanogaster. Evolution, 38(5), 1004—1010. JSTOR. https://doi.org/10.2307/2408434

variation. Given the age-dependent weighting of the force of natural selection on age-specific survival probabilities, the results of these experiments had to be of this kind, *or the core idea of declining forces of natural selection as the primary cause of aging would have been falsified.* Unlike cell-molecular theories of aging, the evolutionary theory of aging made strong falsifiable predictions that put the standing of the theory at risk. Furthermore, those predictions were corroborated in repeated and replicated experiments from different laboratories.

Let us, however, return to the findings of Lansing (e.g., 1948) and Comfort (1953), which also employed late ages of first reproduction, but did not observe an increase in longevity. All of Lansing's rotifer cultures died off, both those reproduced at early ages and those reproduced at later ages. Subsequent rotifer research has featured cultures that did not die off. Thus Lansing's findings seem to be due chiefly to bad laboratory culture technique. Comfort's experiment, with *D. subobscura,* featured inbred lines that lacked sufficient genetic variation to evolve in any direction. In the absence of sufficient standing genetic variation, there is no expectation for any sustained experimental evolution over the course of 10 to 20 generations. In such experimental evolution designs, the experimenter must wait for mutation to generate the genetic variation before selection can produce an evolutionary response (e.g., Lenski et al., 1991).

As we will show repeatedly over the course of subsequent chapters, the results of Rose and Charlesworth (1981b), Luckinbill et al. (1984), and Rose (1984b) have been reproduced many times over. The chief requirements are standing genetic variation and sustained changes in the first age of reproduction of evolving laboratory populations. Thus these strong-inference publications provided the primary experimental paradigm for the last 40 years of research on the evolution of aging (vid. Rauser et al., 2009).

In addition, these experiments provided evidence that the evolutionary postponement of aging was associated with reduced early fecundity, in keeping with the negative genetic correlations between early fecundity and longevity published by Rose and Charlesworth (1981a). This result supported antagonistic pleiotropy as a viable evolutionary genetic mechanism for the evolution of aging. Although concerns were raised that the observed results were the consequence not of a strong trade-off but of reproductive restraint (Partridge, 1987), later work by Service (1989) demonstrated that the divergence between early-reared and late-reared selection regimes was not contingent on their overall reproductive activity, as the same results were found in cohorts of females compared as lifelong virgins. Such virgins have greatly reduced fecundity. Thus the delayed-reproduction paradigm supported the antagonistic pleiotropy mechanism for the evolution of aging, within the theoretical context developed primarily by Charlesworth (e.g., 1980).

In overview, formal mathematical development of the evolutionary theory of aging from 1966 to 1980 was matched with the publication of a robust experimental evolution paradigm for testing and refining that theory from 1980 to 1984. From 1984 onward, then, it is fair to say that the evolutionary biology of aging has been a robust subfield within evolutionary biology as a whole. A less settled question, in the 1980s, was whether or not it could match reductionist cell-molecular theories of aging with respect to parsing the physiological machinery of aging.

References and further reading

Charlesworth, B. (1980). *Evolution in age-structured populations*. Cambridge: Cambridge University Press.
Comfort, A. (1953). Absence of a Lansing effect in *Drosophila subobscura*. *Nature, 172*(4367), 83–84. https://doi.org/10.1038/172083a0

Edney, E. B., & Gill, R. W. (1968). Evolution of senescence and specific longevity. *Nature, 220*(5164), 281—282. https://doi.org/10.1038/220281a0

Hamilton, W. D. (1966). The moulding of senescence by natural selection. *Journal of Theoretical Biology, 12*(1), 12—45. https://doi.org/10.1016/0022-5193(66)90184-6

Lansing, A. I. (1948). Evidence for aging as a consequence of growth cessation. *Proceedings of the National Academy of Sciences of the United States of America, 34*(6), 304—310.

Lenski, R. E., Rose, M. R., Simpson, S. C., & Tadler, S. C. (1991). Long-term experimental evolution in *Escherichia coli*. I. Adaptation and divergence during 2,000 generations. *The American Naturalist, 138*(6), 1315—1341. https://doi.org/10.1086/285289

Luckinbill, L. S., Arking, R., Clare, M. J., Cirocco, W. C., & Buck, S. A. (1984). Selection for delayed senescence in *Drosophila melanogaster*. *Evolution, 38*(5), 996—1003. https://doi.org/10.2307/2408433. JSTOR.

Medawar, P. B. (1946). Old age and natural death. *Modern Quarterly, 1*, 30—56.

Medawar, P. B. (1952). *An unsolved problem of biology*. H.K. Lewis and Co.

Mueller, L. D., Rauser, C. L., & Rose, M. R. (2011). *Does aging stop?* Oxford: Oxford University Press.

Partridge, L. (1987). Is accelerated senescence a cost of reproduction? *Functional Ecology, 1*(4), 317—320. https://doi.org/10.2307/2389786. JSTOR.

Rauser, C., Mueller, L., Travisano, M., & Rose, M. (2009). 18. Evolution of aging and late life. In T. Garland, & M. Rose (Eds.), *Experimental evolution: Concepts, methods, and applications of selection experiments* (pp. 551—584). Berkeley: University of California Press. https://doi.org/10.1525/9780520944473-019

Rose, M. R. (1984). Laboratory evolution of postponed senescence in *Drosophila melanogaster*. *Evolution, 38*(5), 1004—1010. https://doi.org/10.2307/2408434. JSTOR.

Rose, M., & Charlesworth, B. (1980). A test of evolutionary theories of senescence. *Nature, 287*(5778), 141—142. https://doi.org/10.1038/287141a0

Rose, M. R., & Charlesworth, B. (1981a). Genetics of life history in Drosophila melanogaster. I. Sib analysis of adult females. *Genetics, 97*(1), 173—186.

Rose, M. R., & Charlesworth, B. (1981b). Genetics of life history in *Drosophila melanogaster*. II. Exploratory selection experiments. *Genetics, 97*(1), 187—196.

Rose, M. R., Dorey, M. L., Coyle, A. M., & Service, P. M. (1984). The morphology of postponed senescence in *Drosophila melanogaster*. *Canadian Journal of Zoology, 62*(8), 1576—1580. https://doi.org/10.1139/z84-230

Service, P. M. (1989). The effect of mating status on lifespan, egg laying, and starvation resistance in *Drosophila melanogaster* in relation to selection on longevity. *Journal of Insect Physiology, 35*(5), 447—452. https://doi.org/10.1016/0022-1910(89)90120-0

Wattiaux, J. M. (1968a). Cumulative parental age effects in Drosophila subobscura. *Evolution, 22*(2), 406—421. https://doi.org/10.1111/j.1558-5646.1968.tb05908.x

Wattiaux, J. M. (1968b). Parental age effects in *Drosophila pseudoobscura*. *Experimental Gerontology, 3*(1), 55—61. https://doi.org/10.1016/0531-5565(68)90056-9

Williams, G. C. (1957). Pleiotropy, natural selection, and the evolution of senescence. *Evolution, 11*(4), 398—411. https://doi.org/10.2307/2406060. JSTOR.

CHAPTER TWENTY THREE

1977—1988: Characterization of *Caenorhabditis elegans* mutants with extended lifespan

The standard paradigm

In discussing the empirical status of the evolutionary biology of aging, thus far we have largely focused on empirical studies of aging using metazoans, predominantly those with experiments on Drosophila lines. A major problem with this research was that it was based on presuppositions about the genetics of natural populations, but it did not characterize the details of those genetics. In Drosophila evolutionary genetics generally, it was well-known by the 1980s that there was an abundance of molecular genetic variation at individual loci in wild populations, but the functional importance of such locus-by-locus variation for characters like survival and reproduction, and thus Darwinian fitness, was not known (vid. Lewontin, 1974).

For the purpose of examining the molecular genetic foundations of the evolution of aging, the exact nature of longevity mutants at this time remained elusive. Though Maynard Smith (1958) had shown that *ovariless* Drosophila mutants exhibited increased lifespan, there were no cases of delayed or slowed aging mutant stocks that could be readily cultured as homozygotes in the laboratory. Presumably, if a single gene were solely responsible for the onset of aging once maturity is reached, the subsequent mutation of such a gene could potentially lead to an indefinitely extended life span. Such mutations would helpfully reconcile simple cell-molecular hypotheses with the evolutionary biology of aging, if they existed.

The practical challenge was the creation of laboratory strains that had individual mutants with extended lifespan, but without the problem of total infertility exhibited by Maynard Smith's (1958) *ovariless* mutant stocks. Mutants of large-effect with reduced lifespan had been known since the time of Pearl (1922), but their relationship with the genetic variation underpinning lifespan in outbred populations was unclear. Such mutations are strongly selected against in natural populations, as they are typically associated with reduced Darwinian fitness. Thus the field needed extended-lifespan mutants

Conceptual Breakthroughs in The Evolutionary Biology of Aging
ISBN: 978-0-12-821545-6
https://doi.org/10.1016/B978-0-12-821545-6.00026-1

in hand, in order to proceed from the DNA sequences of individual genes to patterns of aging.

The conceptual breakthrough

Klass (1977) as well as Johnson (1982, 1990) found and characterized a *C. elegans* mutant of the *age-1* gene that had extended lifespan but could also be cultured normally in the laboratory. Relative to their control and ancestral wild-type, N2 (Bristol), the hermaphroditic mutants at this locus across four separate replicates had a 65%–110% increase in mean longevity (Johnson, 1990) while exhibiting a traditional mortality curve, only delayed. This longevity increase was preserved even as temperature and nutrition were varied.

Impact: 7

There has been some controversy about whether or not the *age-1(hx546)* mutant has reduced fertility (Johnson & Friedman, 1988), but its extended lifespan has been established beyond reasonable doubt. Two mutant strains totally lacking any active PI3K (the *age-1* gene product), nonsense alleles *mg44* and *m333* that terminate upstream of the kinase domain, were later shown to live up to 10 times the normal lifespan: mean lifespans of 130–170 days, versus 17 days for near-isogenic N2 controls (Ayyadevara et al., 2008). Second-generation homozygotes for the truncated-PI3K alleles were extremely long-lived and were totally sterile.

The overall importance of the introduction of *C. elegans* mutants with reliably extended lifespan is that it gave geneticists in the 1980s two very different kinds of genetic systems that featured extended lifespan: lab-evolved Drosophila and mutant nematodes. It is important to be clear, however, that the provenance of these two systems was quite different. The Drosophila lab cultures that evolved increased lifespan were genetically outbred, and highly variable in their molecular genetics and quantitative genetics. The nematode mutants were created from long-frozen samples isolated from the wild. The source strains for these mutants were not adapted to laboratory conditions. Furthermore, the nematode stocks in which they were maintained were usually homozygous, at least prior to controlled genetic crossing. This made the study of the nematode longevity mutants relatively straightforward using conventional genetic tools. But such nematode stocks have population genetics that are fundamentally different from those of normally outbreeding species, like the vast majority of insect and vertebrate species. This contrast would prove fateful for the subsequent development of the field up until the advent of genomics circa 2000.

References and further reading

Ayyadevara, S., Alla, R., Thaden, J. J., & Shmookler Reis, R. J. (2008). Remarkable longevity and stress resistance of nematode PI3K-null mutants. *Aging Cell, 7*(1), 13—22. https://doi.org/10.1111/j.1474-9726.2007.00348.x. Epub 2007 Nov 7. PMID: 17996009.

Friedman, D. B., & Johnson, T. E. (1988). A mutation in the age-1 gene in *Caenorhabditis elegans* lengthens life and reduces hermaphrodite fertility. *Genetics, 118*(1), 75—86. https://doi.org/10.1093/genetics/118.1.75

Johnson, T. E. (1990). Increased life-span of age-1 mutants in *Caenorhabditis elegans* and lower gompertz rate of aging. *Science, 249*(4971), 908—912 (JSTOR).

Johnson, T. E., & Wood, W. B. (1982). Genetic analysis of life-span in Caenorhabditis elegans. *Proceedings of the National Academy of Sciences of the United States of America, 79*(21), 6603—6607.

Johnson, T. E., Wu, D. Q., Tedesco, P., Dames, S., & Vaupel, J. W. (2001). Age-specific demographic profiles of longevity mutants in *Caenorhabditis elegans* show segmental effects. *The Journals of Gerontology. Series A, Biological Sciences and Medical Sciences, 56*, B331—B339. https://doi.org/10.1093/gerona/56.8.B331

Klass, M. R. (1977). Aging in the nematode *Caenorhabditis elegans*: Major biological and environmental factors influencing life span. *Mechanisms of Ageing and Development, 6*, 413—429. https://doi.org/10.1016/0047-6374(77)90043-4

Klass, M. R. (1983). A method for the isolation of longevity mutants in the nematode *Caenorhabditis elegans* and initial results. *Mechanisms of Ageing and Development, 22*(3—4), 279—286. https://doi.org/10.1016/0047-6374(83)90082-9

Lewontin, R. C. (1974). *The genetic basis of evolutionary change*. New York: Columbia University Press.

Maynard Smith, J. (1958). The effects of temperature and of egg-laying on the longevity of *Drosophila subobscura*. *Journal of Experimental Biology, 35*(4), 832—842.

CHAPTER TWENTY FOUR

1982—85: Further mathematical characterization of evolution with antagonistic pleiotropy

The standard paradigm

The most explicit mathematical treatments of the evolutionary genetics of aging by Charlesworth (Charlesworth, 1980; Charlesworth & Williamson, 1975) focused on the case of mutation accumulation. In particular, Charlesworth showed that progressive mutation accumulation with age is expected to provide abundant and selectable genetic variation for aging-related characters. There are two reasons why this type of evolutionary genetic mechanism had appeal to theoreticians like Charlesworth.

First, selection-mutation balance is a long-standing and important idea in evolutionary theory. Evolutionary theorists have invoked it to explain a wide range of biological phenomena, from the maintenance of genetic diseases that are recessive (chiefly or solely expressed in homozygotes for the disease allele) to the maintenance of genetic variation for quantitative characters (e.g., Lande, 1975). For many evolutionary geneticists, such selection-mutation balance is the chief source of genetic variation segregating in natural populations.

Second, the specific type of mutation accumulation invoked by Medawar (1952) and Charlesworth (e.g., 1980) is mathematically elegant. In developing theory for this type of evolutionary mechanism, each age-class can be considered separately. This makes the relationship between adult age and the amount of standing genetic variation simpler to analyze analytically. However, it should be noted that the core assumption of a complete lack of pleiotropic effects across age-classes is extreme. In the case of Mendelian genetic variants of large effect, a multiplicity of effects across phenotypes and ages is generic, not rare. Yet because so little was known, circa 1980, about patterns of pleiotropy across ages among those alleles that segregate at high frequencies, alleles that do not have the large effects of classic Mendelian alleles, there was no clear reason not to make the mathematically convenient assumption that pleiotropy did not arise.

Conceptual Breakthroughs in The Evolutionary Biology of Aging
ISBN: 978-0-12-821545-6
https://doi.org/10.1016/B978-0-12-821545-6.00020-0

Quantitative-genetic (e.g., Rose & Charlesworth, 1981a) and selection (e.g., Rose & Charlesworth, 1981b) experiments revealed selectable genetic variation for aging, but without any evidence of mutation accumulation. This raised the question of the relationship between antagonistic pleiotropy and the maintenance of genetic variance for aging-related characters. If pleiotropy is commonplace, then mutation accumulation can't be the predominant or sole evolutionary genetic mechanism underpinning the evolution of aging.

The conceptual breakthrough

Over three articles, Rose (1982, 1983, 1985) showed that antagonistic pleiotropy often leads to the maintenance of genetic polymorphisms. At selective equilibrium, these genetic polymorphisms have little additive genetic variance for fitness itself. But they nonetheless feature abundant selectable (i.e., additive genetic) variation for individual life-history characters.

For example, Rose (1985) analyzed the impact of antagonistic pleiotropy in the context of age-structured populations. Consider the case of antagonistic pleiotropic genetic effects on age-specific fecundities, with a single locus having alleles A_1 and A_2, and age-specific fecundities, $m(x)$ at ages x_a and x_b. Here, any beneficial effect due to homozygosity is represented by $h_a f_a$ and $h_b f_b$ and any deleterious effect by f_a and f_b for each day, respectively. Rose (1985) precluded heterozygote overdominance for the individual fitness parameters, since that is a trivial way to enable a theoretical model to predict the maintenance of genetic variation. The age-specific fecundities can then be expressed as the following:

	$A_1 A_2$	$A_1 A_2$	$A_2 A_2$
$m_{ij}(x_a)$	$m_{ij}(x_a) - f_a$	$m_{ij}(x_a)$	$m(x_a) + h_a f_a$
$m_{ij}(x_b)$	$m_{ij}(x_b) + h_b f_b$	$m_{ij}(x_b)$	$m(x_b) - f_b$

In order to maintain genetic polymorphisms, the following expressions must be sufficiently large.

$$e^{-r_{11}x_a}\, l(x_a)f_a - e^{-r_{11}x_b}\, l(x_b)h_b f_b \tag{24.1}$$

As well as

$$e^{-r_{22}x_b}\, l(x_b)f_b - e^{-r_{22}x_a}\, l(x_a)h_a f_a \tag{24.2}$$

Here $l(x)$ represents the age-specific survivorship, as before, and the dominance parameters are h_a and h_b. When these dominance parameters are small, it is more likely that polymorphism will be maintained. Naturally, this is only one of the cases considered by Rose (1985), but it illustrates his general finding that the maintenance of genetic polymorphisms is common with antagonistic pleiotropy in the absence of overdominance.

From the standpoint of testing whether or not antagonistic pleiotropy has shaped the evolutionary genetics of natural populations, the maintenance of genetic polymorphism involving *some cases* of alleles with antagonistic pleiotropy implies that it will be possible to detect the presence of antagonistic pleiotropy in a population's evolutionary history. In effect, those particular loci with segregating alleles that show antagonistic pleiotropy indicates the occurrence of other alleles with antagonistic pleiotropy that were driven to fixation by selection.

Given this demonstration that antagonistic pleiotropy can sustain genetic polymorphisms, then the possibility that it has played a role of some type in the evolution of aging can be testing by studying standing genetic variation. For example, when a negative genetic correlation is detected between longevity and early fecundity in an outbred population, as occurred in the study of Rose and Charlesworth (1981a), then that is evidence that antagonistic pleiotropy has shaped the evolution of longevity in that population. Similarly, when the evolution of increased average lifespan occurs during experimental evolution, we can test for antagonistic pleiotropy by looking for reductions in early functional characters, like development time or early fecundity.

Thus the relatively simple theoretical analyses provided by Rose (e.g., 1985) imply that when antagonistic pleiotropy has shaped the evolution of aging, then there are straightforward experimental tests which should reveal its signature. If there are no such patterns in standing genetic variation, then it is reasonable to infer that antagonistic pleiotropy has not shaped the evolution of aging. Once again, the evolutionary genetic theory of aging provides opportunities for strong-inference tests of its alternative component hypotheses.

Impact: 7

Templeton (1980) and Abugov (1986) also pursued this line of theoretical work, as have still later theorists. One of its more general implications
pertains to the longstanding issue of the maintenance of genetic variation in
natural populations generally. Few cases of heterozygote superiority at single
loci of large effect have been found, excepting that of sickle-cell anemia
polymorphisms in malarial regions. Notably, however, even that case is an
example of antagonistic pleiotropy; alleles that foster sickling reduce respira-
tory efficiency, but they increase resistance to malarial infection. If there are
numerous loci of small effect that exhibit antagonistic pleiotropy, thanks to
recessive deleterious effects, then there could be an abundance of selectable
genetic variation for functional characters segregating in most natural populations. Such variation provides an explanation for the successes of agricultural breeding as well as the widespread success of experimental evolution
that starts with diploid sexual populations that aren't inbred (vid. Garland
& Rose, 2009). In contrast, it provides evidence for why inbred populations
are generally deficient with respect to some functional attributes, though not
all: inbreeding will make most loci homozygous, making recessive deleterious allele effects more prominent.

Thus the antagonistic pleiotropy hypothesis that has been particularly
emphasized in discussions of the evolution of aging leads to a variety of
important corollaries for the scientific study of functional genetic variation
in general, beyond the context of aging specifically. Firstly, it provides an
explanation for the prevalence of selectable genetic variation affecting functional characters in outbred diploid populations. Secondly, the antagonistic
pleiotropy hypothesis helps explain the manifold deficiencies of inbred derivatives of those populations. In these respects, the evolutionary biology of
aging has contributed significantly to the study of the genetic foundations of
evolution across a range of functional characters.

References and further reading

Abugov, R. (1986). Genetics of darwinian fitness. III. A generalized approach to age struc-
tured selection and life history. *Journal of Theoretical Biology, 122*(3), 311–323. https://
doi.org/10.1016/S0022-5193(86)80123-0
Charlesworth, B. (1980). *Evolution in age-structured populations.* Cambridge: Cambridge University Press.
Charlesworth, B., & Williamson, J. A. (1975). The probability of survival of a mutant gene in
an age-structured population and implications for the evolution of life-histories. *Genetics
Research, 26*(1), 1–10. https://doi.org/10.1017/S0016672300015792

Garland, T., & Rose, M. R. (Eds.). (2009). *Experimental evolution: Concepts, methods, and applications of selection experiments*. University of California Press.

Lande, R. (1975). The maintenance of genetic variability by mutation in a polygenic character with linked loci. *Genetical Research, 26*(3), 221—235. https://doi.org/10.1017/S0016672300016037

Lande, R. (1980). The genetic covariance between characters maintained by pleiotropic mutations. *Genetics, 94*(1), 203—215. https://doi.org/10.1093/genetics/94.1.203

Medawar, P. B. (1952). *An unsolved problem of biology*. H.K. Lewis and Co.

Rose, M. R. (1982). Antagonistic pleiotropy, dominance, and genetic variation. *Heredity, 48*(1), 63—78. https://doi.org/10.1038/hdy.1982.7

Rose, M. R. (1983). Further models of selection with antagonistic pleiotropy. In H. I. Freedman, & C. Strobeck (Eds.), *Population biology. Lecture notes in biomathematics* (Vol 52). Berlin, Heidelberg: Springer. https://doi.org/10.1007/978-3-642-87893-0_7

Rose, M. R. (1985). Life history evolution with antagonistic pleiotropy and overlapping generations. *Theoretical Population Biology, 28*(3), 342—358. https://doi.org/10.1016/0040-5809(85)90034-6

Rose, M. R., & Charlesworth, B. (1981a). Genetics of life history in Drosophila melanogaster. I. Sib analysis of adult females. *Genetics, 97*(1), 173—186.

Rose, M. R., & Charlesworth, B. (1981b). Genetics of life history in *Drosophila melanogaster*. II. Exploratory selection experiments. *Genetics, 97*(1), 187—196.

Templeton, A. R. (1980). The evolution of life histories under pleiotropic constraints and r-selection. *Theoretical Population Biology, 18*(2), 279—289. https://doi.org/10.1016/0040-5809(80)90053-2

CHAPTER TWENTY FIVE

1984: Genetic covariation is shifted to positive values by inbreeding

The standard paradigm

A central methodological problem in biology is the widespread use of inbred stocks in laboratory experiments. This is not a difficulty when experiments concern attributes of limited functional significance, at least in a laboratory setting, such as flower color in plants or eye color in mammals. For such characters, inbred and mutant laboratory "strains" allow straightforward genetic analysis using a variety of techniques that have been in wide use since the 1910s.

But the study of life-history characters, like viability or longevity, is potentially undermined through the use of inbred stocks, especially when the wild populations from which such stocks are derived do not normally inbreed. Deriving inbred populations from such outbred populations has two consistent and pernicious effects: (1) reduced or otherwise "depressed" functional characters, also known as "inbreeding depression"; and (2) high levels of genetic differentiation *between* inbred lineages in conjunction with reduced levels of genetic variation *within* each such lineage. This is a well-known empirical effect in the quantitative genetic literature (e.g., Falconer, 1981, pp. 1–133), with well-developed formal theory that explains it. However, it is easily instantiated using "pure bred" dog breeds. Such breeds are distinctively different from each other, genetically depauperate within breeds, and generally subject to a wide range of health problems, from hip dysplasia to deafness to blindness. Large wild populations of the wolves from which domesticated dogs were derived are nothing like pure bred dogs, both phenotypically and genetically.

This problem is exacerbated in the biomedical context. Humans normally outbreed, and very few suffer from the kinds of major genetic problems that are characteristic of lab rodent strains that have been heavily inbred to the point of genome-wide homozygosity. This is perhaps easily understood from the contrast between wild wolves and purebred lap dogs.

Conceptual Breakthroughs in The Evolutionary Biology of Aging
ISBN: 978-0-12-821545-6
https://doi.org/10.1016/B978-0-12-821545-6.00025-X

But this problem is completely general to biological research. In the wild, most Drosophila species are highly outbred, and contain extensive genetic variation. In most labs, their Drosophila stocks are either accidently or deliberately inbred. This makes them fully parallel, at the evolutionary genetic level, to the problems of purebred dogs. They are subject to *both* inbreeding depression *and* strong between-strain genetic differentiation.

A common result in experiments with Drosophila stocks and similar genetic model species, such as mice, is thus positive genetic covariation among life-history characters. That is, some lines generally have greater viability, fecundity, and longevity compared to other lines, as illustrated by experiments from the Giesel lab (Giesel, 1979; Giesel et al., 1982; Giesel & Zettler, 1980). But these are predictable results of varying levels of inbreeding depression among the randomly differentiated inbred lines used in such experiments. As the level of inbreeding depression will vary randomly between inbred lines, it can be expected to generate consistent differences between these lines with respect to characters like viability, early fecundity, and longevity. Such consistent differences will in turn generate positive genetic covariances among life-history characters, obliterating the patterns of genetic covariation, and thus pleiotropy, characteristic of the ancestral wild populations from which they were derived.

The conceptual breakthrough

The aforementioned patterns were well-known to animal breeders, but they were not generally understood by those who performed experiments on the genetic variation underpinning the evolution of aging. In order to demonstrate this potential artifact directly, Rose (1984a) deliberately created inbred derivatives of the outbred population in which he and Charlesworth had previously found evidence for antagonistic pleiotropy (e.g., Rose, 1984b; Rose & Charlesworth, 1981a). These newly derived inbred lines then showed a tendency toward positive covariation among life-history characters, as observed in Table 25.1. This was explained by Rose in terms of variation in the extent to which generally deleterious alleles rose in frequency due to genetic drift, among inbred lines. As a result of the stochasticity of genetic drift, inbreeding within established lines could be responsible for the inadvertent changes in sign values of genetic correlations, namely, from negative to positive values.

Thus, Rose (1984a) argued, positive genetic variation among life-history characters in inbred lines does not constitute evidence against antagonistic

Table 25.1 Mean correlations for adult female life-history characters over inbred lines of *D. Melanogaster*.

	Fecundity days 1—5	Fecundity days 6—10	Fecundity days 11—15	Longevity
Fecundity days 1—5	—	0.62	−0.11	0.01
Fecundity days 6—10	0.62	—	0.34	0.50
Fecundity days 11—15	−0.11	0.34	—	0.23
Longevity	0.01	0.50	0.23	—

Data from Rose, M. R. (1984a). Genetic covariation in Drosophila life history: Untangling the data. *The American Naturalist, 123*(4), 565—569. https://doi.org/10.1086/284222.

pleiotropy contributing to the evolution of aging in normally outbreeding populations. More broadly, results from experiments utilizing inbreeding in this fashion to address the genetics of outbred aging are likely to be misleading.

Impact: 6

While this analysis of the impact of inbreeding helped to clear up confusion regarding the evolutionary biology of aging among evolutionary geneticists, it chiefly provided a conceptual cleansing, rather than direct progress. If more evolutionary geneticists had been aware of the animal breeding literature, the experiment would have been unnecessary.

But inbred stocks are still commonly used in gerontological studies, even by some schools of population genetics, such as those who use the DGRP (Mackay et al., 2012). Thus it could be argued that the long-term impact of this work was limited. Most biologists continue to use inbred material in the study of life-history and other functional characters, including patterns of aging. Much gerontological research has thereby been undermined, *especially* in its attempts to apply biomedical findings obtained from inbred rodents to the vast majority of humans, who are rarely inbred.

References and further reading

Bell, G. (1984a). Measuring the cost of reproduction. I. The correlation structure of the life table of a plankton rotifer. *Evolution, 38*(2), 300—313. https://doi.org/10.1111/j.1558-5646.1984.tb00289.x

Curtsinger, J. W. (2019). Fecundity for free? Enhanced oviposition in longevous populations of *Drosophila melanogaster*. *Biogerontology, 20*(4), 397—404. https://doi.org/10.1007/s10522-018-09791-1

Falconer, D. S. (1981). *Introduction to quantitative genetics* (2nd ed.). London: Longman Group Ltd.

Giesel, J. T. (1979). Genetic co-variation of survivorship and other fitness indices in *Drosophila melanogaster*. *Experimental Gerontology, 14*(6), 323—328. https://doi.org/10.1016/0531-5565(79)90044-5

Giesel, J. T., Murphy, P. A., & Manlove, M. N. (1982). The influence of temperature on genetic interrelationships of life history traits in a population of *Drosophila melanogaster*: What tangled data sets we weave. *The American Naturalist, 119*(4), 464—479.

Giesel, J. T., & Zettler, E. E. (1980). Genetic correlations of life historical parameters and certain fitness indices in *Drosophila melanogaster*: R m, rs, diet breadth. *Oecologia, 47*(3), 299—302. https://doi.org/10.1007/BF00398520

Khazaeli, A. A., & Curtsinger, J. W. (2010). Life history variation in an artificially selected population of *Drosophila melanogaster*: Pleiotropy, superflies, and age-specific adaptation. *Evolution, 64*(12), 3409—3416. https://doi.org/10.1111/j.1558-5646.2010.01139.x

Mackay, T. F. C., Richards, S., Stone, E. A., Barbadilla, A., Ayroles, J. F., Zhu, D., Casillas, S., Han, Y., Magwire, M. M., Cridland, J. M., Richardson, M. F., Anholt, R. R. H., Barrón, M., Bess, C., Blankenburg, K. P., Carbone, M. A., Castellano, D., Chaboub, L., Duncan, L., ... Gibbs, R. A. (2012). The *Drosophila melanogaster* genetic reference panel. *Nature, 482*(7384), 173—178. https://doi.org/10.1038/nature10811

Rose, M. R. (1984a). Genetic covariation in Drosophila life history: Untangling the data. *The American Naturalist, 123*(4), 565—569. https://doi.org/10.1086/284222

Rose, M. R. (1984b). Laboratory evolution of postponed senescence in *Drosophila melanogaster*. *Evolution, 38*(5), 1004—1010. https://doi.org/10.2307/2408434. JSTOR.

Rose, M. R., & Charlesworth, B. (1981a). Genetics of life history in Drosophila melanogaster. I. Sib analysis of adult females. *Genetics, 97*(1), 173—186.

Rose, M. R., & Charlesworth, B. (1981b). Genetics of life history in *Drosophila melanogaster*. II. Exploratory selection experiments. *Genetics, 97*(1), 187—196.

CHAPTER TWENTY SIX

1984: Direct demonstration of nonaging in fissile species

The standard paradigm

Qualitative empirical findings sometimes provide strong-inference tests of major theories, but those cases are rare. In physics, one of the most famous examples of this is Galileo using the orbits of Jupiter's moons. Contrary to the geocentric theories that then prevailed among Aristotelian physicists, the moons of Jupiter obviously orbit Jupiter, not the Sun, when those orbits are tracked using telescopes. Contrary to Popperian scientific practice (e.g., Popper, 1959), but in keeping with Kuhn's historiography of actual scientific practice (e.g., Kuhn, 1962), the Aristotelian astrophysicists solved the problem of this strong refutation by refusing to look through telescopes at Jupiter's moons.

In biology, the patterns of fossil differentiation across geological strata are easily explained in terms of Darwin's evolutionary tree theory, but difficult to explain using special creation by a deity. For example, fossils in higher strata resemble current species more than fossils from deeper strata. Once large taxonomic groups go extinct in deep strata, they do not reappear in shallow strata, one famous example of this being the trilobites, the predominant large animals of the Paleozoic, but entirely absent after the start of the Mesozoic. It is to the credit of biology, as a scientific field, that findings like these firmly established Darwinian evolution as a foundational theory for all of biology.

A collection of anecdotes and maximum-longevity records had long supported the idea that some fissile species do not age (reviewed in Comfort, 1979). Of the documented taxa, certain phyla had inconclusive records of senescence, such as *Porifera* (sponges) and *nemerteans* (ribbon worms). Those with more robust evidence for the absence of aging, such as paratomical organisms like *Hydra,* actiniarians, and aelosomatids, nonetheless were not firmly established as nonaging. Among the issues impeding the general acceptance of the phenomenon of nonaging among fissile species was variation in survival patterns found among different laboratory cultures,

Conceptual Breakthroughs in The Evolutionary Biology of Aging
ISBN: 978-0-12-821545-6
https://doi.org/10.1016/B978-0-12-821545-6.00049-2

including patterns of abrupt death associated with failures of culture conditions (vid. Comfort, 1979). Thus, the phenomenon of nonaging was still not widely accepted in the early 1980s.

Yet the validity of the evolutionary theory of aging was at stake on this particular empirical question. For if aging is solely due to declines in the forces of natural selection after the onset of reproduction, then species that do not exhibit such declines in these forces must not show systematic and sustained increases in mortality. This is an absolute prediction of the evolutionary theory, qualified only to the extent that it applies only to species which exhibit fissile reproduction with enough symmetry that there is no discernible "parent" which can be distinguished from "offspring." This qualification is not determined solely by whether reproduction involves cellular fission or not. Asymmetrical fission, such as occurs in brewer's yeast *Saccharomyces cerevisiae*, can still feature a "mother cell" and a "daughter cell". On the other hand, multicellular organisms that reproduce vegetatively may feature enough symmetry of division that there is no discernible difference between products of the vegetative spread.

The question remained, in the early 1980s, whether this pattern was sufficiently consistent in contrasts between species that, for example, reproduced using eggs versus species that reproduced by some type of splitting.

The conceptual breakthrough

Bell (1984) compared the demographic patterns of age-specific survival among small aquatic invertebrate species, some fissile and some not. Of the six species he studied, four were strictly ovigerous (from the genera *Philodina*, *Platyias*, *Cypridopsis*, and *Daphnia*) and two were fissile oligochaetes (*Pristina aequiseta* and *Aelosoma tenebrarum*). The main focus of the experiment was to compare decreases in the rate of survival of individual experimental cohorts *between* the group of species that reproduce vegetatively *with* those species that do not reproduce vegetatively.

Traditional nonevolutionary explanations for senescence, which make no stipulations with respect to patterns of selection, imply that there should be no distinction between the two groups. This would hold especially in species with similar cell biology, such as aquatic invertebrates like those studied by Bell (1984). But the evolutionary theory of aging based on the forces of natural selection requires that the rate of survival among adults *must* decrease in the ovigerous species. The fissile organisms by contrast should *not* exhibit any sustained deterioration in survival rate with adult age, so long as good culture conditions are maintained, and the products of vegetative reproduction are sufficiently similar physiologically.

As the evolutionary theory of aging requires, the Bell experiments demonstrated that those species that reproduced by fission did *not* show an acceleration in mortality rates with age. Those that did not reproduce by fission, as predicted, *did* show such an acceleration. Bell concluded that only evolutionary theories of aging had scientific credibility. This was firm evidence that the existence of senescence is not a physiological inevitability. Instead, it is the result of natural selection shaping age-specific rates of death and reproduction according to the forces of natural selection.

Impact: 9

This study provided a strong-inference test (vid. Platt, 1964) of the validity of evolutionary versus nonevolutionary theories of aging, regardless of the specific genetic or biochemical mechanisms hypothesized by either type of theory. It was the first unequivocal landmark study of its kind, setting the standard against which all subsequent studies of the presence or absence of aging should be measured.

Mention should be made of a subsequent publication from Martinez (1998), which firmly established the nonaging status of fissile *Hydra* kept in good culture conditions. Like the sea anemone anecdote given by Comfort (1979), Martinez has been able to maintain individual *Hydra* indefinitely in his aquaria.

Unfortunately, it should be stated that cell-molecular theories of aging have not been adjusted to reflect these profound findings. Their aging experiments all involve species that have the same essential cell-molecular features: telomeres, lysosomes, free-radical damage, and so on. Yet some of these species have demographic aging, while others do not, in the patterns predicted by the evolutionary theory of aging. There is no physiological necessity to biological aging whatsoever. The process is entirely contingent on age-specific patterns of selection, a point that subsequent work on patterns of mortality plateaus would make even more apparent in the 1990s.

Evidently, the findings of experiments of this kind are as abhorrent for cell-molecular biologists as telescopic studies of the orbits of the moons of Jupiter were for Aristotelian astrophysicists in the 17th Century. That may be why they generally ignore them.

References and further reading

Bell, G. (1984). Evolutionary and nonevolutionary theories of senescence. *The American Naturalist, 124*(4), 600–603.

Comfort, A. (1979). *The biology of senescence* (3rd ed.). Churchill Livingstone Edinburgh and London.

Kuhn, T. S. (1962). *The structure of scientific revolutions*. University of Chicago Press. https://books.google.com/books?id=a7DaAAAAMAAJ

Martínez, D. E. (1998). Mortality patterns suggest lack of senescence in Hydra. *Experimental Gerontology, 33*(3), 217–225. https://doi.org/10.1016/S0531-5565(97)00113-7

Platt, J. R. (1964). Strong Inference: Certain systematic methods of scientific thinking may produce much more rapid progress than others. *Science, 146*(3642), 347–353. https://doi.org/10.1126/science.146.3642.347

Popper, K. R. (1959). *The logic of scientific discovery*. New York, NY: Basic Books.

CHAPTER TWENTY SEVEN

1989: Additional experiments support antagonistic pleiotropy

The standard paradigm

Science cannot entirely rest on a small collection of experiments conducted in a small number of settings. In biology, the problem of generalizing across species is well-known, and sometimes acute. There are plant species, such as the evening primrose studied by Hugo de Vries, which prolifically generate new reproductively isolated strains due to large-scale chromosome rearrangements (Mayr, 1982). But such cases turn out to involve unusual cytogenetic machinery which is rarely found among other species, especially animals. In the vast majority of cases, "instant speciation" is not the primary manner in which new species arise (Mayr, 1982).

With the first publications providing evidence in support of antagonistic pleiotropy as a genetic mechanism underlying the evolution of aging (e.g., Rose & Charlesworth, 1981a,b), the question of the ubiquity of such phenomena as negative genetic correlations between life-history characters became one of active controversy. From a historical standpoint, it is fortunate that Andrew Clark gave a useful summary of the state of debate in the mid-1980s in his 1987 article (Clark, 1987), outlining the issues surrounding genetic correlations and alternative population genetic mechanisms for the generation of such correlations.

A useful point that Clark raises is the relationship between models of mutation-selection balance, as described by Charlesworth (1980) and Lande (1980), and assumptions about patterns of pleiotropy in mutation-selection models focused on stabilizing selection. Among Clark's more interesting contributions to defining the state of conventional thinking about the evolutionary genetics of aging, was the following: "in these (mutation-selection) simulations, negative genetic correlations are generated, but mean longevity decreases only when there are pleiotropic mutations when the effects of mutations on the two phenotypic attributes correlate negatively" (Clark, 1987).

Conceptual Breakthroughs in The Evolutionary Biology of Aging
ISBN: 978-0-12-821545-6
https://doi.org/10.1016/B978-0-12-821545-6.00013-3

The conceptual breakthrough

Research with populations other than laboratory Drosophila was published in the later 1980s, some of it supporting the existence of antagonistic pleiotropy (e.g., Butlin & Day, 1989). In the 1989 Butlin and Day article, the model organism was the seaweed fly, *Coelopa frigida*, and the focal characters studied were wing size (serving as a surrogate for body size) and fecundity. They found a particular karyotype that sacrificed early fecundity for greater net fecundity in total. This inversion was also responsible for an overall increase in longevity, a similar genetic trade-off pattern to that observed by Rose (1984b) in a different species.

By the end of the 1980s, the genetic mechanism of antagonistic pleiotropy was fairly well established as a factor underlying life-history evolution generally. Indeed, this pattern of pleiotropy made the evolution of aging effectively inextricable from the evolution of life history generally. Thus, the invocation of antagonistic pleiotropy as a genetic mechanism for the evolution of aging was no longer controversial.

Impact: 6

The accumulating evidence in support of antagonistic pleiotropy demonstrated that it was not a phenomenon limited to laboratory Drosophila, but a common genetic mechanism to be considered in the evolutionary biology of aging generally. Indeed, the theoretical work of Rose (1982, 1983, 1985) suggested that it might be a widespread factor in the maintenance of genetic variation affecting life-history characters of all kinds. However, nothing about such theory or its supporting experimental data precluded the possibility of genetic variation being maintained by mutation-selection. And therefore, such mathematical and empirical findings suggesting the importance of antagonistic pleiotropy did *not* show that mutation accumulation couldn't also contribute to the evolution of aging.

References and further reading

Butlin, R. K., Collins, P. M., & Day, T. H. (1984). The effect of larval density on an inversion polymorphism in the seaweed fly Coelopa frigida. *Heredity, 52*(3), 415–423. https://doi.org/10.1038/hdy.1984.49

Butlin, R. K., & Day, T. H. (1989). Environmental correlates of inversion frequencies in natural populations of seaweed flies (Coelopa frigida). *Heredity, 62*(2), 223–232. https://doi.org/10.1038/hdy.1989.32

Charlesworth, B. (1980). *Evolution in age-structured populations.* Cambridge: Cambridge University Press.

Clark, A. G. (1987). Senescence and the genetic-correlation hang-up. *The American Naturalist, 129*(6), 932—940. https://doi.org/10.1086/284686

Lande, R. (1980). The genetic covariance between characters maintained by pleiotropic mutations. *Genetics, 94*, 203—215.

Mayr, E. (1982). Speciation and macroevolution. *Evolution, 36*(6), 1119—1132. https://doi.org/10.1111/j.1558-5646.1982.tb05483.x

Rose, M. R. (1982). Antagonistic pleiotropy, dominance, and genetic variation. *Heredity, 48*(1), 63—78. https://doi.org/10.1038/hdy.1982.7

Rose, M. R. (1983). Further models of selection with antagonistic pleiotropy. In H. I. Freedman, & C. Strobeck (Eds.), *Population biology* (Vol. 52, pp. 47—53). Springer Berlin Heidelberg. https://doi.org/10.1007/978-3-642-87893-0_7

Rose, M. R. (1984b). Laboratory evolution of postponed senescence in *Drosophila melanogaster*. *Evolution, 38*(5), 1004—1010. https://doi.org/10.2307/2408434. JSTOR.

Rose, M. R. (1985). Life history evolution with antagonistic pleiotropy and overlapping generations. *Theoretical Population Biology, 28*(3), 342—358. https://doi.org/10.1016/0040-5809(85)90034-6

Rose, M. R., & Charlesworth, B. (1981a). Genetics of life history in *Drosophila melanogaster*. I. Sib analysis of adult females. *Genetics, 97*(1), 173—186.

Rose, M. R., & Charlesworth, B. (1981b). Genetics of life history in *Drosophila melanogaster*. II. Exploratory selection experiments. *Genetics, 97*(1), 187—196.

CHAPTER TWENTY EIGHT

1985: Genotype-by-environment interaction shown for aging

The standard paradigm

A second major methodological problem with laboratory experiments in biology is the problem of genotype-by-environment interactions (usually abbreviated as GxE). GxE is such a commonplace phenomenon for aging-related characters that it should be assumed as an experimental design risk. In more than 40 years of experimental research on aging, we have found it to bedevil experimental design and interpretation so often that we regard it as the greatest hazard facing reproducibility in this field, so long as inbreeding is evaded (Inbreeding, as we have already discussed can entirely vitiate experiments on aging. But it is an obvious experimental design flaw. GxE can arise cryptically and in surprising ways).

Though it can be a recondite hazard in experimentation, GxE is an entirely predictable result of assaying the aging-related characters of populations that have not previously evolved under the culture conditions of the assay. For example, wild-caught rodents that are studied in adult cohorts fed laboratory rodent chow are undergoing aging under entirely novel conditions. Since aging is an age-dependent loss of adaptation, feeding an animal a novel foodstuff to which it is not adapted is to impose a second cause of failure of adaptation. Furthermore, comparisons of wild-caught animals from multiple species in the laboratory will produce patterns of genetic and environmental variation, including covariation that must depend in part on the degree of preadaptation of those species to that laboratory environment. Notice that the degree of preadaptation can in principle, and often in practice, vary among animal species. This type of complication is entirely predictable in advance. But the sad fact of many biological fields, from comparative biology to much of gerontology, is that this problem is usually neglected.

But GxE is a more severe problem for experiments on aging than the previous paragraph suggests. We have found that it obscures, if not wholly undermines, experimental interpretation even when we, as experimenters, were determined to avoid it. Perhaps the best explanation we can give for

Conceptual Breakthroughs in The Evolutionary Biology of Aging
ISBN: 978-0-12-821545-6
https://doi.org/10.1016/B978-0-12-821545-6.00053-4
117

the insidious difficulties of GxE in studies of life-history characters is that life-history characters numerically determine fitness. If there is any opportunity for a species to evolve patterns of environmental responsiveness of life-history characters that can increase fitness, then it is likely for a species to do so. The literature on "reaction norms" in life-history research, reviewed well by Stearns (1992), makes this point abundantly.

Perhaps the experimental context where the challenges of GxE for aging research are most severe is the study of patterns of genetic covariation among life-history characters. Specifically, when life histories are assayed or selected on in novel environments, one possibility is that there are allelic variants at multiple loci which generally favor or disfavor life history, across all ages, in such novel conditions. This would tend to bias patterns of genetic covariation toward positive values, away from negative values that would provide evidence for antagonistic pleiotropy.

This concern is exemplified by the rotifer studies of Bell (1984a, 1984b), whose experimental organisms had been recently introduced to the laboratory conditions they were assessed in. Bell (1984a, 1984b) assessed the fecundity and longevity patterns of rotifers but did not allow them sufficient time to evolutionarily adapt to laboratory conditions, prior to his assay.

This raises the possibility of artificial effects from such a novel environment, relative to populations that have reached an evolutionary equilibrium. The intuition is that there could be segregating genetic variation that affects general suitedness to the new environment, thereby enhancing life-history characters without any antagonism among these effects. Such segregating genetic variation would then bias genetic correlations among functional characters toward positive values, though not necessarily elevating them past the zero value. This effect could have created the positive correlations among life-history characters that were found in Bell's (1984a, 1984b) experiments on aquatic invertebrates, as shown in Figs. 28.1 and 28.2.

The conceptual breakthrough

Service and Rose (1985) and Luckinbill and Clare (1985) sought to gauge the significance of novel environments for genetic correlations, using their respective populations of Drosophila. Each experiment used comparisons of *Drosophila melanogaster* results obtained with entirely novel assay environments versus results obtained with environments to which the study populations had adapted. Their goal was to determine if genetic correlations would be biased toward positive values due to the shift to a novel environment.

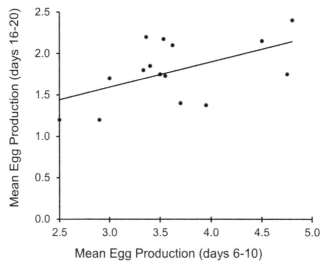

Figure 28.1 The sum of eggs produced during days 16–20 is plotted against the sum of eggs produced during days 6–10 in clones under laboratory conditions. The results provide ostensible evidence against cost of reproduction theories. *Data from Bell, G. (1984a). Measuring the cost of reproduction. I. The correlation structure of the life table of a plankton rotifer. Evolution, 38(2), 300–313. https://doi.org/10.1111/j.1558-5646.1984. tb00289.x.*

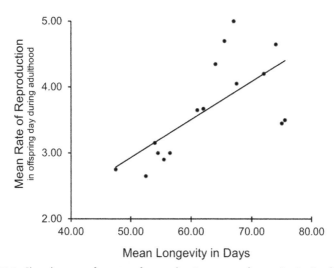

Figure 28.2 Clonal means for rate of reproduction versus longevity in *Daphnia pulex* reared in a laboratory environment. A strong positive correlation is evident between the two variables under the experimental conditions. *Data from Bell, G. (1984b). Measuring the cost of reproduction. II. The correlation structure of the life tables of five freshwater invertebrates. Evolution, 38(2), 314–326. https://doi.org/10.1111/j.1558-5646. 1984.tb00290.x.*

Service and Rose (1985) compared genetic correlations under two assay conditions, one deliberately contrived to be as different as practicable from the normal assay conditions. The first assay was conducted using the normal conditions, with food created from banana-agar and karo syrups, at a temperature of 25°C. The second assay used a novel diet, *Instant Drosophila Medium-Blue*, with an altered temperature of 15.5°C. To reduce any initial maternal effects, two generations of standard rearing at a controlled density of 30 flies per vial were employed before introducing one of the assay groups to the novel treatment conditions. To assess the impact of the novel conditions, phenotypic assays tracked mean fecundity (egg output per 24 h) and mean starvation time (in hours), and the genetic correlations (r_A) between the two characters were estimated.

When the novel-assay group was compared to the normal-assay group, Service and Rose (1985) found that the additive genetic correlation between the two functional phenotypes was significantly increased under novel assay conditions. This was a direct experimental demonstration that genotype-by-environment interactions could significantly bias genetic correlations toward positive values.

Luckinbill and Clare (1985) also examined this question by comparing lines of *Drosophila melanogaster* that had evolved with either early-reproduction or late-reproduction regimens over a number of generations. They manipulated larval density and assessed the impact of that manipulation on adult longevity, relative to the selection regime previously imposed. They found that varying larval density resulted in changes in the magnitude of observable differentiation between these two types of population, directly establishing a genotype-by-environment interaction for longevity under conditions of varying larval rearing density.

Impact: 7

It should be noted that no organism subject to experimental conditioning is wholly free of handling issues. However, the findings of Service and Rose (1985) and Luckinbill and Clare (1985) show how improperly designed experiments can bias estimates of genetic correlations upward, thereby obscuring the presence of antagonistic pleiotropy, as well as impeding the efficacy of selection itself. These findings were corroborated almost a decade later by Leroi et al. (1994).

However, while *evolutionary biologists* have become aware of this general problem of experimental design, it is rarely considered in the experimental work of *other* biologists. Even some evolutionary geneticists doing research on the experimental evolution of *Escherichia coli* (e.g., their papers on detecting genetic trade-offs during adaptation to temperature variations, such as Bennett & Lenski, 1997) fail to appreciate that their tests for evolutionary genetic trade-offs among functional characters are often compromised by their use of experimental paradigms that feature novel environments.

Among gerontologists, this type of artifact is rarely allowed for. Thus, there is still controversy about the presence or absence of genetic trade-offs associated with mutants that extend longevity in *C. elegans* (vid. Van Voorhies et al., 2006), even in stocks of the nematode that have never adapted to the assay conditions that are used. The presence of trade-offs in *C. elegans* mutants is discussed further in Chapter 48.

Overall, despite the salience of this work on genotype-by-environment interaction as a source of artifacts, it has had relatively little impact on subsequent work across the subdisciplines of biology, including those parts of gerontology uninformed by evolutionary research.

References and further reading

Bell, G. (1984a). Measuring the cost of reproduction. I. The correlation structure of the life table of a plankton rotifer. *Evolution, 38*(2), 300−313. https://doi.org/10.1111/j.1558-5646.1984.tb00289.x

Bell, G. (1984b). Measuring the cost of reproduction. II. The correlation structure of the life tables of five freshwater invertebrates. *Evolution, 38*(2), 314−326. https://doi.org/10.1111/j.1558-5646.1984.tb00290.x

Bennett, A. F., & Lenski, R. E. (1997). Evolutionary adaptation to temperature. VI. phenotypic acclimation and its evolution in escherichia coli. *Evolution, 51*(1), 36−44. https://doi.org/10.1111/j.1558-5646.1997.tb02386.x

Leroi, A. M., Chippindale, A. K., & Rose, M. R. (1994). Long-term laboratory evolution of a genetic life-history trade-off in *Drosophila melanogaster*. 1. The role of genotype-by-environment interaction. *Evolution, 48*(4), 1244−1257. https://doi.org/10.2307/2410382. JSTOR.

Luckinbill, L. S., & Clare, M. J. (1985). Selection for life span in *Drosophila melanogaster*. *Heredity, 55*(1), 9. https://doi.org/10.1038/hdy.1985.66

Service, P. M., & Rose, M. R. (1985). Genetic covariation among life-history components: The effect of novel environments. *Evolution, 39*(4), 943−945. https://doi.org/10.2307/2408694. JSTOR.

Service, P. M., Hutchinson, E. W., MacKinley, M. D., & Rose, M. R. (1985). Resistance to environmental stress in *Drosophila melanogaster* selected for postponed senescence. *Physiological Zoology, 58*(4), 380−389.

Stearns, S. C. (1992). *The evolution of life histories*. Oxford University Press.

Van Voorhies, W. A., Curtsinger, J. W., & Rose, M. R. (2006). Do longevity mutants always show trade-offs? *Experimental Gerontology, 41*(10), 1055−1058. https://doi.org/10.1016/j.exger.2006.05.006

CHAPTER TWENTY NINE

1985—onward: Evolutionary physiology of aging

The standard paradigm

Since the 1960s, conventional biogerontologists have assumed that aging is due to basic cell-molecular processes of physiological deterioration, such as free-radical damage, somatic mutation, mitochondrial deterioration, etc. This, however, has generally been an unexamined presupposition. The possibility that aging might also involve organ-system or whole-organism physiology is a common view among geriatricians, who often deal with problems of organ failure and the like in their patients, even when such aging-associated disorders have no apparent or demonstrated connection to free radicals or mutations.

But there are substantial problems with inferring physiological causes from chronological associations with such demographic characters as declining age-specific survival probabilities. Comfort (1979, pp. 34—35) outlined some of the concerns associated with using physiological measurements to infer the mechanistic causes of aging. Biological characters may be spuriously correlated with chronological age but have nothing to do with the actual causes of deterioration underlying biological senescence. This necessitates an effort to distinguish between physiological characters that are merely correlated with adult age versus those physiological characters that causally determine deteriorating survival and reproduction of the organism in question.

This predicament is exemplified by the work of Ganetzky and Flanagan (1978), who endeavored to find "landmarks" of senescence. Inspired by work performed by Burcombe (1972), who demonstrated age-dependent decreases in the enzymatic activity of alcohol dehydrogenase (ADH) in *Drosphila*, Ganetzky and Flanagan (1978) sought to use this biochemical feature as an indicator of aging. Two strains of *Drosophila melanogaster*, Oregon-R (OR) and Canton-S (CS), were chosen because they were known to differ in male longevity by approximately 20 days. It was presumed that the longer-lived OR males should have appreciably more ADH activity than the shorter-lived males of CS, when compared at corresponding ages.

Conceptual Breakthroughs in The Evolutionary Biology of Aging
ISBN: 978-0-12-821545-6
https://doi.org/10.1016/B978-0-12-821545-6.00056-X

The results however were inverted: the CS males demonstrated significantly more ADH activity into late ages relative to their longer-lived counterparts. This provided evidence that physiological characters may decline with age but have no reliable causal relationship with the aging process itself. This suggests that chronological associations between physiological characters and declining survival rates in any one cohort or population will not necessarily indicate that those physiological characters reliably underpin biological aging itself. Instead, the most successful assay from the Ganetzky and Flanagan (1978) study involved locomotor activity. By tracking the time it took for flies tapped to the bottom of a vial to rise toward the top of a vial, they found a physiological character that was more robustly associated with senescence, at least in the strains that they studied.

The stage was set for a more evolutionary approach to physiology. Fortunately, the laboratory evolution of populations with different patterns of aging provided material with which to examine the physiological machinery controlling aging.

The conceptual breakthrough

Service et al. (1985) performed an experiment comparable to that of Ganetzky and Flanagan (1978), exploring the physiological differences between the longer-lived populations and their shorter-lived ancestral controls from Rose (1984b). Furthermore, they used five populations of each type, as well as dozens of flies per assay, an experimental scale much greater than had been the norm in physiological studies of that time (Indeed, physiological experimentation remains notably poor with respect to replication and scale). Table 29.1 shows that the longer-lived populations were not universally superior in all physiological characteristics (Fig. 29.1).

It is also worthy of note that the patterns of some of these physiological characters as biomarkers of aging is inconsistent. For instance, desiccation resistance declined with age for both females and males, and increased desiccation resistance was found among longer-lived populations. But starvation resistance, which was increased in longer-lived flies, had a pattern of increasing with adult age. Explanations for such paradoxical patterns are explored further in Chapter 31.

Service (1989) examined similar physiological characters to those of his previous studies, but focused on the physiological impact of mating status to determine if the differences between the long-lived and short-lived populations were conserved when the flies in question were virgins. On the

Table 29.1 — Differentiated characters observed in longer-lived "O" populations of *D. melanogaster*, relative to shorter-lived "B" populations.

Enhanced characters with increased lifespan	Depressed characters with increased lifespan
Later fecundity	Early fecundity
Starvation resistance (all ages)	Early ovary weight
Desiccation resistance (all ages)	Early metabolic rate
Ethanol resistance (all ages)	Early locomotor activity
Later locomotor activity	
Lipid content	

From Rose, M. R. (1984b). Laboratory evolution of postponed senescence in *Drosophila melanogaster*. *Evolution, 38*(5), 1004—1010. JSTOR. https://doi.org/10.2307/2408434; Service, P. M., Hutchinson, E. W., MacKinley, M. D., & Rose, M. R. (1985). Resistance to environmental stress in *Drosophila melanogaster* selected for postponed senescence. *Physiological Zoology, 58*(4), 380—389, and Service, P. M. (1987). Physiological mechanisms of increased stress resistance in *Drosophila melanogaster* selected for postponed senescence. *Physiological Zoology, 60*(3), 321—326.

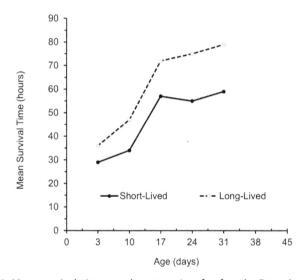

Figure 29.1 Mean suvival times under starvation for female *D. melanogaster* from longer lived (in dashes) versus shorter lived (solid) stocks. Flies from longer-lived stocks tend to survive starvation longer at most ages. *Data from Service, P. M., Hutchinson, E. W., MacKinley, M. D., & Rose, M. R. (1985). Resistance to environmental stress in Drosophila melanogaster selected for postponed senescence.* Physiological Zoology, 58(4), 380—389.

whole, while the mating status of both males and females tended to have a significant impact on physiological characters *within* groups, the patterns of differentiation between groups with different longevities were conserved.

Impact: 7

The evolutionary physiology of aging was first studied by Service et al. (1985) and Service (1987, 1989), as well as Graves et al. (1988), and proved instrumental to extending a bridge between the evolutionary explanation of aging and the study of the physiological mechanisms of aging. More specifically, such work suggested that the evolution of aging was underpinned by organismal characters like stress resistance and locomotor performance, rather than being determined entirely by cell-molecular phenomena like molecular damage (Fig. 29.2).

Findings like these were important foundation stones for the burgeoning field of evolutionary physiology, a new discipline connecting evolutionary biology to physiology as a whole. The field of evolutionary physiology has gone on to become a major component of physiological research in general.

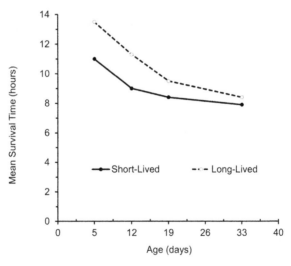

Figure 29.2 Mean survival times under desiccation for female *D. melanogaster* from longer lived (in dashes) versus shorter lived (solid) stocks. Though both exhibit declining age-specific desiccation resistance, flies from longer-lived stocks tend to survive desiccation longer at most ages. *Data from Service, P. M., Hutchinson, E. W., MacKinley, M. D., & Rose, M. R. (1985). Resistance to environmental stress in* Drosophila melanogaster *selected for postponed senescence.* Physiological Zoology, 58(4), 380–389.

References and further reading

Burcombe, J. V. (1972). Changes in enzyme levels during ageing in *Drosophila melanogaster*. *Mechanisms of Ageing and Development, 1*, 213—225.

Comfort, A. (1979). *The biology of senescence* (3rd ed.). Edinburgh and London: Churchill Livingstone.

Ganetzky, B., & Flanagan, J. R. (1978). On the relationship between senescence and age-related changes in two wild-type strains of *Drosophila melanogaster*. *Experimental Gerontology, 13*(3—4), 189—196. https://doi.org/10.1016/0531-5565(78)90012-8

Graves, J. L., Luckinbill, L. S., & Nichols, A. (1988). Flight duration and wing beat frequency in long- and short-lived *Drosophila melanogaster*. *Journal of Insect Physiology, 34*(11), 1021—1026. https://doi.org/10.1016/0022-1910(88)90201-6

Rose, M. R. (1984b). Laboratory evolution of postponed senescence in *Drosophila melanogaster*. *Evolution, 38*(5), 1004—1010. https://doi.org/10.2307/2408434. JSTOR.

Service, P. M. (1987). Physiological mechanisms of increased stress resistance in *Drosophila melanogaster* selected for postponed senescence. *Physiological Zoology, 60*(3), 321—326.

Service, P. M. (1989). The effect of mating status on lifespan, egg laying, and starvation resistance in *Drosophila melanogaster* in relation to selection on longevity. *Journal of Insect Physiology, 35*(5), 447—452. https://doi.org/10.1016/0022-1910(89)90120-0

Service, P. M., Hutchinson, E. W., MacKinley, M. D., & Rose, M. R. (1985). Resistance to environmental stress in *Drosophila melanogaster* selected for postponed senescence. *Physiological Zoology, 58*(4), 380—389.

CHAPTER THIRTY

1987: Accelerated senescence explained in terms of mutation accumulation with inbreeding depression

The standard paradigm

With the growing evidence that aging was best understood in terms of evolutionary biology, further attention was given to the role that the common evolutionary phenomenon of inbreeding depression might play in aging, a phenomenon first directly demonstrated for longevity by Maynard Smith et al. (1955). This was a concern with respect to early experiments on accelerated aging like those of Sokal (1970), whose work with flour beetles hinted at the potential of the experimental tests proposed by Edney and Gill (1968). Since inbreeding depression is a general effect across functional characters, small population sizes in the work of Sokal might have caused the acceleration of senescence, not the shift in the timing of the decline in the forces of natural selection.

Most experiments that attempted to create experimental lines of accelerated senescence at this time did so with a focus on genetic theories. Citing the works of Medawar (1952) and Williams (1957), the aim was largely to test the aging theories of mutation accumulation versus antagonistic pleiotropy.

An alternative to this line of inquiry were experiments focused instead on ecological theories, which sought to explain how environmental conditions might influence the patterns of life history characteristics. One ecological question related to senescence, and with limited empirical support, was what environmental conditions would favor evolution to create two alternate forms of reproductive strategies: semelparous versus iteroparous. A semelparous life history is characterized as having only one reproductive event or bout, followed by near immediate death. Iteroparous life-histories permit multiple instances of reproduction and a slower decline in rates of survival with age. The framework of MacArthur and Wilson's (1967) r/K selection theory distinguished between natural selection focusing more on the Malthusian parameter r or the carrying capacity, K, the population size at which population growth stops. The common intuition among ecologists was that environments that favor semelparity will have an abundance of resources, favor juvenile survival over adult longevity, and be more prone to poor adult survival. By contrast, an environment that would favor the iteroparous strategy would have a relatively large adult census, possibly with lower juvenile survival due to more limited resources for development. A comparison between populations with these two differing reproductive strategies could, in theory, be relevant to the relative merits of alternative genetic theories of senescence.

The conceptual breakthrough

The aim of Mueller (1987) was to use the ecological framework of r/K selection to create accelerated development populations of *Drosophila melanogaster* using an experimental evolution regime that favored increased values of r, coupled with an experimental evolution regime that sustained populations near their carrying capacity, and thus fostered K selection. The experimental setup of Mueller (1987) consisted of six populations, three of which were subjected to an "accelerated r regimen" using exclusively young adults in discrete generations. The other three populations were given a "longer-lived K regimen", with overlapping generations that allowed reproduction by adults of all ages. This experimental design of Mueller (1987) featured notable improvements over preceding acceleration studies, in that all the lines were derived from one common ancestral population and there were replicate populations within treatment groups.

The parameter by which aging was measured was age-specific female fecundity. For the first 4 weeks of adult life there were no differences between the fecundity rates of flies derived from the two selection regimes, but during the fifth week of fecundity comparisons, the r flies had a 47% $-$83% reduction in fecundity compared to the K flies. Bridging ecological theory with the genetic theories of senescence, Mueller (1987) suggested that this pattern of fecundity differentiation could be explained with mutation accumulation theory, while providing little to no evidence for antagonistic pleiotropy. As Bierbaum et al. (1989) noted, during the early generations of selection the r flies did not exhibit an increase in early fecundity or a loss of late fecundity or longevity. But after $\sim$ 100 generations of r

selection, late fecundity was depressed while early fecundity remained unaltered. Mueller proposed that the eventual decline in later fecundity was due to a buildup of deleterious mutations with effects confined to late ages within the *r* populations. That is to say, his interpretation was that mutation accumulation depressed later life fecundity in the *r* populations.

Impact: 7

What holds this score down a bit is the likelihood that maintaining an intentionally small population, as Mueller (1987) noted, increased the likelihood of genetic drift resulting in fixation of weakly deleterious alleles. On this interpretation, the fact that the *r* lines appeared to be more susceptible to the impacts of inbreeding depression lends caution to aging studies employing populations with very small effective population sizes.

On the other hand, it could be concluded that evolving populations in nature are intermittently subject to low population sizes, conditions which will specifically foster the process of mutation accumulation having deleterious effects especially at late ages. In this respect, then, this study from Mueller (1987) emphasizes the dependence of evolutionary genetic mechanisms of aging on the ecological circumstances facing populations.

Finally, note that most laboratory populations are likely to have smaller population sizes than wild populations of the same species. This will tend to bias the evolutionary genetics of their aging toward mutation accumulation, and away from antagonistic pleiotropy in the laboratory. This is another source of observational bias that impedes the detection of antagonistic pleiotropy in laboratory experiments.

References and further reading

Bierbaum, T. J., Mueller, L. D., & Ayala, F. J. (1989). Density-dependent evolution of life-history traits in *Drosophila melanogaster*. *Evolution, 43*(2), 382—392. https://doi.org/10.1111/j.1558-5646.1989.tb04234.x

Edney, E. B., & Gill, R. W. (1968). Evolution of senescence and specific longevity. *Nature, 220*(5164), 281—282. https://doi.org/10.1038/220281a0

MacArthur, R. H., & Wilson, E. O. (1967). *The theory of island biogeography*. Princeton University Press.

Maynard Smith, J., Clarke, J. M., & Hollingsworth, M. J. (1955). The expression of hybrid vigour in *Drosophila subobscura*. *Proceedings of the Royal Society of London. Series B—Biological Sciences, 144*(915), 159—171. https://doi.org/10.1098/rspb.1955.0042

Medawar, P. B. (1952). *An unsolved problem of biology*. H.K. Lewis and Co.

Mertz, D. B. (1975). Senescent decline in flour beetle strains selected for early adult fitness. *Physiological Zoology, 48*(1), 1—23. https://doi.org/10.1086/physzool.48.1.30155634

Mueller, L. (1987). Evolution of accelerated senescence in laboratory populations of Drosophila. *Proceedings of the National Academy of Sciences of the United States of America, 84*, 1974—1977. https://doi.org/10.1073/pnas.84.7.1974

Pianka, E. R. (1970). On r- and K-selection. *The American Naturalist, 104*(940), 592—597. https://doi.org/10.1086/282697

Sokal, R. R. (1970). Senescence and genetic load: Evidence from Tribolium. *Science (New York, N.Y.), 167*(3926), 1733—1734. https://doi.org/10.1126/science.167.3926.1733

Williams, G. C. (1957). Pleiotropy, natural selection, and the evolution of senescence. *Evolution, 11*(4), 398—411. https://doi.org/10.2307/2406060. JSTOR.

CHAPTER THIRTY ONE

1988: Reverse evolution of aging

The standard paradigm

If the evolutionary theory of aging is correct, and especially if antagonistic pleiotropy is an important genetic mechanism underlying the evolution of aging, then faster rates of aging should readily evolve by reverse selection for early reproduction in outbred populations that have first evolved slower rates of aging as a result of delayed first reproduction. This prediction is a natural extension of the original proposals of Edney and Gill (1968), and in keeping with the findings of Sokal (1970). The difference is that the postponed-aging stocks created by, for example, Rose (1984b) were maintained with enough attention to population size that inbreeding depression would be unlikely to play a role in subsequent reacceleration of aging (cf. Mueller, 1987).

In addition, functional traits that rapidly reverse-evolve in such experimental paradigms would be implicated as being involved in an antagonistic pleiotropy syndrome. By contrast, functional traits that are enhanced by selection for later reproduction, but do *not* reverse evolve back to their ancestral state, could conceivably be free of antagonistic pleiotropy. The latter pattern, indeed, would suggest the possibility that some aspects of aging evolve by mutation accumulation. Thus, the sequence of evolutionary increases in lifespan and other functional characters, followed by reverse evolution of lifespan, provides potentially useful tests of both genetic mechanisms for the evolution of aging, antagonistic pleiotropy and mutation accumulation.

There is, however, a possible complication facing reverse evolution experiments. By the mid-1980s, evolutionary experiments had created multiple long-lived lines of Drosophila (Luckinbill et al., 1984; Rose, 1984b) by strengthening the forces of natural selection at later ages. However, any aging experiments implementing reverse selection for early reproduction require that the loci that control aging were not fixed in the prior course of selection for later reproduction, whether by selection itself or by inbreeding. Otherwise, they would lack selectable and segregating alleles that could respond to reverse selection. However, this potential limiting

Conceptual Breakthroughs in The Evolutionary Biology of Aging
ISBN: 978-0-12-821545-6
https://doi.org/10.1016/B978-0-12-821545-6.00060-1

factor would be revealed by failures of reverse evolution experiments to produce shorter lifespans and increased early fecundity. If such failures don't occur, then reverse evolution would indeed be a powerful tool for evaluating the evolutionary genetics of aging.

The conceptual breakthrough

Service et al. (1988) sought to address some of these questions by employing reverse selection for early reproduction starting from the longer-lived populations developed by Rose (1984b). They studied starvation resistance, desiccation resistance, resistance to low-level ethanol, and early fecundity, along with lifespan itself. Service et al. (1988) found that early fecundity increased as longevity declined, in conformity with antagonistic pleiotropy. Similarly, starvation resistance fell as early fecundity rose, again suggesting the role of antagonistic pleiotropy. By contrast, desiccation and ethanol resistance did not exhibit any statistically significant decline, suggesting a pattern more congruent with the mutation accumulation mechanism for the evolution of aging.

Impact: 8

The findings of Service et al. (1988) suggested that the two commonly discussed genetic mechanisms for the evolution of aging, antagonistic pleiotropy and mutation accumulation, could be at work in the populations that they studied. Their work also suggested the value of reverse selection for discriminating among genetic mechanisms of experimental evolution generally.

These two major findings would inspire further work over subsequent decades, especially the use of reverse selection across the field of Drosophila experimental evolution (e.g., Passananti et al., 2004; Teotonio et al., 2009; Teotónio & Rose, 2000). All told, this work established that outbred populations readily evolve changes in patterns of aging as the age of first reproduction shifts up or down. The theoretical paradigm established by Hamilton (1966) and Charlesworth (e.g., 1980) was thus strongly corroborated. In addition, many of the functional characters associated with aging, from patterns of fecundity to stress resistance, shifted in ways that suggested that they were involved in antagonistic pleiotropy syndromes. This work thus continued to underscore the strength of the antagonistic pleiotropy mechanism as a central mechanism for the evolution of aging, as originally argued by Williams (1957) and analyzed mathematically primarily by Rose (e.g., 1985).

References and further reading

Charlesworth, B. (1980). *Evolution in age-structured populations.* Cambridge: Cambridge University Press.

Edney, E. B., & Gill, R. W. (1968). Evolution of senescence and specific longevity. *Nature, 220*(5164), 281—282. https://doi.org/10.1038/220281a0

Hamilton, W. D. (1966). The moulding of senescence by natural selection. *Journal of Theoretical Biology, 12*(1), 12—45. https://doi.org/10.1016/0022-5193(66)90184-6

Luckinbill, L. S., Arking, R., Clare, M. J., Cirocco, W. C., & Buck, S. A. (1984). Selection for delayed senescence in *Drosophila melanogaster. Evolution, 38*(5), 996—1003. https://doi.org/10.2307/2408433. JSTOR.

Mertz, D. B. (1975). Senescent decline in flour beetle strains selected for early adult fitness. *Physiological Zoology, 48*(1), 1—23. https://doi.org/10.1086/physzool.48.1.30155634

Mueller, L. (1987). Evolution of accelerated senescence in laboratory populations of Drosophila. *Proceedings of the National Academy of Sciences of the United States of America, 84*, 1974—1977. https://doi.org/10.1073/pnas.84.7.1974

Passananti, H. B., Deckert-Cruz, D. J., Chippindale, A. K., Le, B. H., & Roses, M. R. (2004). Reverse evolution of aging in *Drosophila melanogaster.* In *Methuselah flies* (Vols. 1—0, pp. 296—322). WORLD SCIENTIFIC. https://doi.org/10.1142/9789812567222_0027

Rose, M. R. (1984a). Genetic covariation in Drosophila life history: Untangling the data. *The American Naturalist, 123*(4), 565—569. https://doi.org/10.1086/284222

Rose, M. R. (1984b). Laboratory evolution of postponed senescence in *Drosophila melanogaster. Evolution, 38*(5), 1004—1010. https://doi.org/10.2307/2408434. JSTOR.

Rose, M. R. (1985). Life history evolution with antagonistic pleiotropy and overlapping generations. *Theoretical Population Biology, 28*(3), 342—358. https://doi.org/10.1016/0040-5809(85)90034-6

Service, P. M., Hutchinson, E. W., MacKinley, M. D., & Rose, M. R. (1985). Resistance to environmental stress in *Drosophila melanogaster* selected for postponed senescence. *Physiological Zoology, 58*(4), 380—389.

Service, P. M., Hutchinson, E. W., & Rose, M. R. (1988). Multiple genetic mechanisms for the evolution of senescence in *Drosophila melanogaster. Evolution; International Journal of Organic Evolution, 42*(4), 708—716. https://doi.org/10.1111/j.1558-5646.1988.tb02489.x

Sokal, R. R. (1970). Senescence and genetic load: Evidence from Tribolium. *Science (New York, N.Y.), 167*(3926), 1733—1734. https://doi.org/10.1126/science.167.3926.1733

Teotónio, H., Chelo, I. M., Bradić, M., Rose, M. R., & Long, A. D. (2009). Experimental evolution reveals natural selection on standing genetic variation. *Nature Genetics, 41*(2), 251—257. https://doi.org/10.1038/ng.289

Teotónio, H., & Rose, M. R. (2000). Variation in the reversibility of evolution. *Nature, 408*(6811), 463. https://doi.org/10.1038/35044070

Williams, G. C. (1957). Pleiotropy, natural selection, and the evolution of senescence. *Evolution, 11*(4), 398—411. https://doi.org/10.2307/2406060. JSTOR.

CHAPTER THIRTY TWO

1985—88: Genetic analysis of aging in males

The standard paradigm

Most research on the evolution of aging, especially in laboratory Drosophila, has been conducted on females, and rarely on males. Virility, the male analogue of fecundity, is particularly neglected. In females, fecundity is easily measured by counts of eggs. Virility is far more difficult to measure, as it often requires the use of multiple females, and assays of the paternity of their offspring.

Given that the definition of aging employed in the evolutionary biology of aging is the persistent decline in adult age-specific fitness components due to endogenous physiological deterioration, such declines in virility are necessarily of interest to the field. Many nonaging related issues might impede successful coupling. Assays of male longevity do not require procedures that are particularly different from those for females, but without estimates of virility, these longevity assays only offer half the information necessary to assess male aging. Yet virility is notably hard to assay.

The conceptual breakthrough

Kosuda (1985) used traditional Drosophila cytogenetic techniques to produce *D. melanogaster* lines with different sets of chromosomes, some heterozygous and some homozygous. Unusually, these lines were studied for their differences in male mating success across a range of adult ages. Surprisingly, the heterogeneity among lines increased with adult age, suggesting the accumulation of increased amounts of genetic variation for virility as a function of age. This result is in keeping with the mutation accumulation genetic mechanism for the evolution of aging. But it should be noted that patterns of genetic variation among lines produced in the manner employed by Kosuda (1985) are generally difficult to interpret. In particular, the maintenance of genetic variation for life-history characters by antagonistic pleiotropy can have complicated effects on overall genetic variance as a function of age

Conceptual Breakthroughs in The Evolutionary Biology of Aging
ISBN: 978-0-12-821545-6
https://doi.org/10.1016/B978-0-12-821545-6.00061-3

(vid. Rose, 1991, p. 74; Hughes & Charlesworth, 1994), making definitive interpretation of results like those of Kosuda (1985) difficult.

Hughes and Clark (1988) studied patterns of genetic variation and covariation among Drosophila chromosomes assayed for their effects on male survival. Among the issues raised in their study was whether mutation accumulation played a role in the evolution of aging. Hughes and Clark (1988) created four sets of recombinant lines from four separate isogenic lines of Drosophila using balancer chromosome methods. Both the initial four isogenic lines and the 53 recombinants were raised and scored daily for their longevity in both sexes and net fecundity. Early fecundity was measured as the "sum of egg production for the first 5 days of posteclosion." Early fecundity and female longevity were only found to be negatively core-lated in one recombinant line (Hughes & Clark, 1988). With all lines pooled, the correlations between these traits were somewhat negative, however.

Impact: 6

Chromosome extraction offers a different perspective on patterns of genetic variation and covariation, whether lines are isogenic or recombinant, compared to studies of quantitative genetic variability within outbred populations (cf. Rose & Charlesworth, 1981a). In particular, the method offers a limited view of the breadth of genetic variation and covariation that characterizes outbred populations with abundant genetic variation. In addition, the alleles making up founding isogenic lines are not a random sample of the genetic variation present in the ancestral population, not least because only those epistatic combinations of alleles that permit line viability and fertility will be present in successfully established isogenic lines. While these studies offered moderate support for the existence of both antagonistic pleiotropy and mutation accumulation, it is open to question whether methods like this are ever sufficiently reliable or powerful in studies of life-history characters.

References and further reading

Hughes, K. A., & Charlesworth, B. (1994). A genetic analysis of senescence in Drosophila. *Nature, 367*(6458), 64–66. https://doi.org/10.1038/367064a0

Hughes, D. M., & Clark, A. G. (1988). Analysis of the genetic structure of life history of *Drosophila melanogaster* using recombinant extracted lines. *Evolution; International Journal of Organic Evolution, 42*(6), 1309–1320. https://doi.org/10.1111/j.1558-5646.1988.tb04190.x

Kosuda, K. (1985). The aging effect on male mating activity in *Drosophila melanogaster*. *Behavior Genetics, 15*(3), 297—303. https://doi.org/10.1007/BF01065984

Rose, M. R. (1991). *Evolutionary biology of aging*. Oxford University Press.

Rose, M. R., & Charlesworth, B. (1981). Genetics of life history in *Drosophila melanogaster*. I. Sib analysis of adult females. *Genetics, 97*(1), 173—186.

CHAPTER THIRTY THREE

1987−1991: Quantitative genetic analysis of how many genes determine aging

The standard paradigm

A longstanding and central question of gerontology has been the mechanistic complexity of aging. Indeed, this very inquiry was first posed by Aristotle. With the collapse in the quality of biological research after Aristotle and his students, exploration of the matter did not entirely resurface until the 1950s. From the speculations of Bidder through the somatic mutation and error catastrophe theories, it appears that gerontologists were fond of promoting singular but ramifying physiological causes for aging.

By contrast, it has been common for evolutionary biologists to allow the possibility that aging could have whatever mechanistic complexity that the genetic variation underlying age-specific life-history characters allowed. In the case of mutation accumulation, evidence of abundant molecular genetic variation among most animal species was well-established by the 1970s. By the 1980s, a common view among theoretical population geneticists was that an abundance of slightly deleterious molecular genetic variation across many loci affected most characters, where this variation was sustained by recurrent mutation. In this context, the mutation accumulation analysis of Charlesworth (e.g., 1980) naturally implied that, to the extent that mutation-selection balance sustained genetic variation for early-life characters like viability during development, then still more such genetic variation at many loci should impinge on aging and its constituent characters.

The case of antagonistic pleiotropy is less obvious, with respect to genetic and mechanistic complexity. The chief reason is that this evolutionary genetic mechanism depends on mutations that have both beneficial and deleterious effects on life-history characters across a range of ages. The frequency at which alleles with antagonistic pleiotropy arise is not as intuitively apparent as the occurrence of merely deleterious alleles. But there is an important contrast between alleles with opposed effects on life-history characters compared to alleles with solely deleterious effects. As shown by Rose

Conceptual Breakthroughs in The Evolutionary Biology of Aging
ISBN: 978-0-12-821545-6
https://doi.org/10.1016/B978-0-12-821545-6.00028-5

141

(e.g., 1985), loci that have allelic variation involving antagonistic pleiotropy will often generate selectively sustained genetic polymorphism. Thus genetic polymorphisms involving antagonistic pleiotropy can build up over time, contributing substantial molecular and quantitative genetic polymorphism affecting aging.

Thus, both mutation accumulation and antagonistic pleiotropy have the potential to produce extensive genetic variation affecting aging, across an indeterminate number of loci. These possibilities inclined evolutionary biologists to emphasize the likelihood that aging is underpinned by a complex array of genes, gene products, and metabolic pathways. But a theoretical likelihood or possibility does not necessarily lead to certain conclusions.

Before the 1980s, cell-molecular biologists had proposed single, merely physiological, mechanisms for aging, but none of these mechanisms had proven notably successful, as we have reviewed above. This left more gerontologists open to the idea that aging arises from multiple, distinct, physiological mechanisms working in concert, as opposed to a single unitary physiological mechanism of aging.

The discovery of "longevity genes" in nematodes (e.g., Friedman & Johnson, 1988; Johnson, 1990) raised the issue of the number of genes that underlie the control of normal aging. The question that this in turn raised was exactly how many genes were involved in the evolution of aging, and what this would mean for the evolutionary explanation of aging (Cutler 1982).

The conceptual breakthrough

Luckinbill et al. (1987) sought to address this question by estimating the number of effective genetic factors that contributed to the evolution of delayed senescence in the longer-lived *Drosophila melanogaster* stocks that they had created (Luckinbill et al., 1984). They used a crossing-technique developed by Sewall Wright, based on changes in patterns of phenotypic variance in sequential generations of hybridization and back-crossing. Unexpectedly their results went against the expected patterns of high polygenicity for given traits, instead ostensibly showing that about one major locus was involved in the differentiation between short- and long-lived stocks.

Luckinbill et al. (1988) continued to address the gene number question but switched to the analysis of chromosomal contributions to aging differentiation among the same stocks from Luckinbill et al. (1984). Using balancer stocks, they were able to show that genes affecting the differentiation of their stocks were to be found on all major chromosomes. Unlike the findings of Luckinbill et al. (1987), which suggested a single genetic factor involved in the evolution of aging, the Luckinbill et al. (1988) study provided evidence of longevity being a largely polygenetic trait.

Hutchinson and Rose (1991) and Hutchinson et al. (1991) also studied segregation indices in similar crosses of the short-lived and long-lived *D. melanogaster* first described in Rose (1984a), using much more replication than that of Luckinbill et al. (1987). They showed that the segregation index used by Luckinbill et al. (1987) was subject to a key mathematical artifact, when its denominator passes through zero. But when the segregation index is inverted, that mathematical artifact is removed, and the index can be used to test critically a null hypothesis of polygenicity. When such inverted segregation-index tests were performed, Hutchinson and Rose (1990) found that segregation index data did *not* support the involvement of only a few genes in the evolution of aging.

Impact: 7

The work of Luckinbill et al. (1988), Hutchinson and Rose (1990,1991), and Hutchinson et al. (1991) was not definitive with respect to determining the number of loci involved in the laboratory evolution of aging. But together, they firmly established that it was unlikely that a single gene controlled the experimental evolution of postponed aging in their Drosophila stocks, contrary to the stunning result of Luckinbill et al. (1987).

In the context of nematode genetics, this finding has been corroborated repeatedly with steadily more powerful technology (e.g., Ayyadevara et al., 2001, 2003; Ebert et al., 1993, 1996). Nematode aging is affected by many loci. The puzzle of genetic and thus mechanistic complexity was settled definitively in the 2010s, as we will discuss below in later chapters.

References and further reading

Ayyadevara, S., Ayyadevara, R., Hou, S., Thaden, J. J., & Reis, R. J. S. (2001). Genetic mapping of quantitative trait loci governing longevity of *Caenorhabditis elegans* in recombinant-inbred progeny of a bergerac-BO × RC301 interstrain cross. *Genetics, 157*(2), 655—666. https://doi.org/10.1093/genetics/157.2.655

Ayyadevara, S., Ayyadevara, R., Vertino, A., Galecki, A., Thaden, J. J., & Reis, R. J. S. (2003). Genetic loci modulating fitness and life span in *Caenorhabditis elegans*: Categorical trait interval mapping in CL2a × bergerac-BO recombinant-inbred worms. *Genetics, 163*(2), 557—570. https://doi.org/10.1093/genetics/163.2.557

Charlesworth, B. (1980). *Evolution in age-structured populations*. Cambridge: Cambridge University Press.

Cutler, R. G. (1982a). Longevity is determined by specific genes: Testing the hypothesis. In R. Adelman, & G. Roth (Eds.), *Testing the theories of aging* (pp. 25—114). Boca Raton, FL: CRC Press.

Ebert, R. H., Cherkasova, V. A., Dennis, R. A., Wu, J. H., Ruggles, S., Perrin, T. E., & Shmookler Reis, R. J. (1993). Longevity-determining genes in *Caenorhabditis elegans*: Chromosomal mapping of multiple noninteractive loci. *Genetics, 135*(4), 1003—1010. https://doi.org/10.1093/genetics/135.4.1003

Ebert, R. H., Shammas, M. A., Sohal, B. H., Sohal, R. S., Egilmez, N. K., Ruggles, S., & Shmookler Reis, R. J. (1996). Defining genes that govern longevity inCaenorhabditis elegans. *Developmental Genetics, 18*(2), 131—143. https://doi.org/10.1002/(SICI)1520-6408(1996)18:2<131::AID-DVG6>3.0.CO;2-A

Friedman, D. B., & Johnson, T. E. (1988). A mutation in the age-1 gene in *Caenorhabditis elegans* lengthens life and reduces hermaphrodite fertility. *Genetics, 118*(1), 75—86. https://doi.org/10.1093/genetics/118.1.75

Hutchinson, E. W., & Rose, M. R. (1990). Quantitative genetic analysis of postponed aging in *Drosophila melanogaster*. *Genetic Effects on Aging, 11*, 66—87.

Hutchinson, E. W., & Rose, M. R. (1991). Quantitative genetics of postponed aging in *Drosophila melanogaster*. I. Analysis of outbred populations. *Genetics, 127*(4), 719—727. https://doi.org/10.1093/genetics/127.4.719

Hutchinson, E. W., Shaw, A. J., & Rose, M. R. (1991). Quantitative genetics of postponed aging in *Drosophila melanogaster*. II. Analysis of selected lines. *Genetics, 127*(4), 729—737. https://doi.org/10.1093/genetics/127.4.729

Johnson, T. E. (1990). Increased life-span of age-1 mutants in *Caenorhabditis elegans* and lower gompertz rate of aging. *Science, 249*(4971), 908—912 (JSTOR).

Luckinbill, L. S., Arking, R., Clare, M. J., Cirocco, W. C., & Buck, S. A. (1984). Selection for delayed senescence in *Drosophila melanogaster*. *Evolution, 38*(5), 996—1003. https://doi.org/10.2307/2408433. JSTOR.

Luckinbill, L. S., Clare, M. J., Krell, W. L., Cirocco, W. C., & Richards, P. A. (1987). Estimating the number of genetic elements that defer senescence in Drosophila. *Evolutionary Ecology, 1*(1), 37—46. https://doi.org/10.1007/BF02067267

Luckinbill, L. S., Graves, J. L., Reed, A. H., & Koetsawang, S. (1988). Localizing genes that defer senescence in *Drosophila melanogaster*. *Heredity, 60*(3), 367. https://doi.org/10.1038/hdy.1988.54

Rose, M. R. (1984a). Genetic covariation in Drosophila life history: Untangling the data. *The American Naturalist, 123*(4), 565—569. https://doi.org/10.1086/284222

Rose, M. R. (1984b). Laboratory evolution of postponed senescence in *Drosophila melanogaster*. *Evolution, 38*(5), 1004—1010. https://doi.org/10.2307/2408434. JSTOR.

Rose, M. R. (1985). Life history evolution with antagonistic pleiotropy and overlapping generations. *Theoretical Population Biology, 28*(3), 342—358. https://doi.org/10.1016/0040-5809(85)90034-6

CHAPTER THIRTY FOUR

1988: Evidence for senescence in the wild

The standard paradigm

The conventional view of evolutionary theorists was that aging should not be readily observable in the wild, leaving aside the spectacularly abrupt aging of semelparous animals and annual plants, immediately after reproduction. This view may have been bolstered by the great difficulty of discerning true aging in a wild setting. For instance, a common practice in field studies assessing aging is to exclude any instances of death due to any serious injury or limitation posed by the environment. The intent is to attribute to aging only those deaths due to endogenous declines of fitness characters, not an external force like predation or disease. In doing so however, the pool of appropriate test subjects in the wild is greatly limited, where few individuals are expected to survive anywhere close to the ages at which laboratory cohorts show obvious deterioration.

Comfort suggested that adult mortality in vertebrates might not decrease with age within wild populations in light of evidence of increased survival with age in wild birds (Comfort, 1979, pp. 142–143). That said, very few vertebrate species terminate their life cycle due to senescence, making efforts to study wild aging all the more difficult. Given the challenge to observe aging in iteroparous vertebrates in the wild, the case for iteroparous invertebrates should prove even more trying, though some successful work was indeed done (Edmondson, 1945a, 1945b).

Semelparous organisms, those that reproduce only once and experience nearly immediate death afterward, are the subject of some debate. Many in the field of gerontology do not consider semelparous organisms to be experiencing senescence in the traditional sense, but a pseudo-form that instead is better characterized as "programmed death" (e.g., Comfort, 1979, p. 140; Kirkwood & Cremer, 1982). From an evolutionary perspective on aging, however, there is nothing about the rapid deterioration of semelparous organisms after reproduction that is profoundly different from supposedly

Conceptual Breakthroughs in The Evolutionary Biology of Aging
ISBN: 978-0-12-821545-6
https://doi.org/10.1016/B978-0-12-821545-6.00038-8

"normal" aging. In fact, given the significant difficulties facing the measurement of aging in the wild, such species may be the most dramatic examples of trade-offs between adult survival and fecundity that can be observed in the wild.

The conceptual breakthrough

Nesse (1988) challenged the notion that ecological sources of mortality act so early that appreciable aging does not occur in the wild. He proposed that, if the conventional view is correct, wild populations should rarely exhibit an increase in death rates after the first age of reproduction. If wild organisms do age, then their death rates should increase after adult maturity due to the physiological impact of aging.

Nesse (1988) used demographic data collected from populations of large vertebrates. From those data, he constructed hypothetical life tables where senescence does not occur and calculated the reproductive advantage associated with the absence of senescence. For some of the species that he analyzed, the observed age-specific patterns of mortality and reproduction did not fit the aging-free hypothetical life tables, suggesting that detectable senescence was in fact present among older individuals in those species.

Impact: 6

Because it is difficult and expensive to study aging in laboratory animals much larger than rodents, Nesse's (1988) demonstration that there might be observable aging among wild populations of large vertebrates provides additional opportunities for studying aging in species ill-suited to laboratory research. In particular, since a major goal among many who study senescence is to gain better insights into human gerontology specifically, large mammals in the wild might be an important window into some aspects of aging in humans, because we ourselves are large mammals that do not live and breed in laboratories, leaving aside some desperate graduate students and postdocs.

References and further reading

Comfort, A. (1979). *The biology of senescence* (3rd ed.). Edinburgh and London: Churchill Livingstone.

Edmondson, W. T. (1945a). Ecological studies of sessile rotatoria, part I: Factors affecting distribution. *Ecological Monographs, 14*(2), 31—66.

Edmondson, W. T. (1945b). Ecological studies of sessile rotatoria, part II: Dynamics of populations and social structures. *Ecological Monographs, 15*(2), 141—172. https://doi.org/10.2307/1948601

Kirkwood, T. B., & Cremer, T. (1982). Cytogerontology since 1881: A reappraisal of August Weismann and a review of modern progress. *Human Genetics, 60*(2), 101—121. https://doi.org/10.1007/bf00569695

Nesse, R. M. (1988). Life table tests of evolutionary theories of senescence. *Experimental Gerontology, 23*(6), 445—453. https://doi.org/10.1016/0531-5565(88)90056-3

1989—onward: Molecular genetic variation at selected loci in the evolution of aging

The standard paradigm

A characteristic challenge to the evolutionary biology of aging from cell-molecular gerontologists is the assertion by them that only their methods can reveal the molecular genetic underpinnings of aging. Beginning in the 1960s with Watson and Crick pushing for a more reductionist approach to biology, biogerontology largely followed suit with a great emphasis being placed on cells and molecules rather than whole-organism research. A natural incentive for this shift is convenience and cost: eliminating large cohorts of laboratory animals saves on physical space, time, and expense, compared to in vitro cell cultures. Cytogerontologists primarily perceived aging through the lens of the somatic cell. For them it was an article of faith that aging was nothing more or less than cell-molecular damage, inefficient repair of that damage, and thus limitations to the proliferation and functioning of such damaged cells.

The conceptual breakthrough

Starting in 1989, evolutionary biologists developed an alternative strategy for determining the molecular foundations of aging, one that did not depend on in vitro cell research. At first, they used simple technologies like protein electrophoresis, but eventually they proceeded to more advanced molecular technologies based on genomics, as will be shown in this and subsequent chapters.

Molecular-genetic assays of Drosophila that had evolved increased longevity were pioneered by Luckinbill et al. (1989). As discussed previously in Chapter 33, Luckinbill et al. (1988b) had concluded that the genetic foundations underlying the experimental evolution of postponed aging are highly polygenic. This finding inspired them to attempt to link individual loci with the evolutionary physiology that their research and that of

Conceptual Breakthroughs in The Evolutionary Biology of Aging
ISBN: 978-0-12-821545-6
https://doi.org/10.1016/B978-0-12-821545-6.00052-2

Service et al. (e.g., 1985) had shown to underpin extended longevity and reproductive lifespan. These physiological characters included starvation resistance, desiccation resistance, and flight endurance (Graves et al., 1988; Service et al., 1985).

The specific aim of Luckinbill et al. (1989) was to focus on candidate loci that could be plausibly associated with such changes in organismal physiology, because of their known roles in intermediary metabolism. In particular, their technology of choice was protein electrophoresis, since it was well-known to effectively characterize at least some of the genetic variation at the loci which coded for the amino acid sequence variation that in turn determined differences in the metabolic pathways that these proteins catalyzed (vid. e.g., Ayala et al., 1972).

Their study found four loci that exhibited significantly differentiated frequencies of electrophoretic variants between short- and long-lived flies. For example, they found differentiation for glucose-6-phosphate dehydrogenase (G6PD). Given the increased flight endurance of their longer-lived flies (Graves et al., 1988), Luckinbill et al. (1989) were able to construct a metabolic narrative whereby an increased synthesis of glycogen, thanks to more active G6PD, would enhance both flight endurance and longevity. This was a pioneering use of experimental evolutionary analysis to connect a functional physiological trait, flight endurance, with underlying molecular machinery involving a well-understood biochemical pathway.

This pioneering effort using protein electrophoresis was continued by Fleming et al. (1993), Tyler et al. (1993), and Deckert-Cruz et al. (1997, 2004). These subsequent studies demonstrated the presence of widespread molecular-genetic differentiation underpinning the experimental evolution of aging in laboratory Drosophila. For example, Fleming et al. (1993) used two-dimensional gel electrophoresis to search for differentiation between the control and longer-lived populations created by Rose (1984b) among proteins produced by about 300 genetic loci that were *not* chosen based on a priori suppositions. Using combinatoric analysis and parsimony trees, Fleming et al. (1993) attempted to estimate the number of loci across the genome that might underpin the experimental evolution of postponed aging. Their estimates of this number suggested that a few percent of the protein-coding loci might control aging in Drosophilia. With a genome size of about 14,000 such loci, that implied that hundreds of loci might be involved in the control of aging in that species. This early estimate would prove to be very important for the genomic research that would follow in the 21st Century.

Tyler et al. (1993) and Deckert-Cruz et al. (1997, 2004) also used one-dimensional protein electrophoresis like Luckinbill et al. (1989), focusing on candidate loci. Two candidates of particular interest were phosphoglucomutase and CuZn-superoxide dismutase, with their electrophoretic allele frequencies strongly associated with the experimental evolution of aging, shifting in a manner suggestive of antagonistic pleiotropy. Some of this work invited criticism (Allikian et al., 2002), but the parallelism among Drosophila species found in the work of Deckert-Cruz et al. (1997, 2004) suggested that the inferred changes in electrophoretic allele frequencies were not likely to be due to genomic hitch-hiking artifacts (cf. Maynard Smith & Haigh, 1974).

Impact: 8

Despite the cytogerontologists' claims of preeminent insight into the molecular foundations of aging, experimentally evolved populations that have different patterns of aging provide excellent material for the study of the molecular foundations of aging. Indeed, because experimental evolutionists can study those molecular foundations of aging with or without a priori candidate mechanisms, it is at least a reasonable proposition to contend that experimental evolutionary genetics is a more powerful foundation for the study of the actual mechanisms of organismal aging than the study of the molecular determinants of in vitro cell proliferation or function.

References and further reading

Allikian, M. J., Deckert-Cruz, D., Rose, M. R., Landis, G. N., & Tower, J. (2002). Doxycycline-induced expression of sense and inverted-repeat constructs modulates phosphogluconate mutase (Pgm) gene expression in adult *Drosophila melanogaster*. *Genome Biology, 3*(5). https://doi.org/10.1186/gb-2002-3-5-research0021. research0021.1.

Ayala, F. J., Powell, J. R., Tracey, M. L., Mourão, C. A., & Pérez-Salas, S. (1972). Enzyme variability in the *Drosophila willistoni* group. IV. genetic variation in natural populations of *Drosophila willistoni*. *Genetics, 70*(1), 113—139. https://doi.org/10.1093/genetics/70.1.113

Deckert-Cruz, D. J., Matzkin, L. M., Graves, J. L., & Rose, M. R. (2004). Electrophoretic analysis of methuselah flies from multiple species. In *Methuselah flies* (Vols. 1—0, pp. 237—248). World Scientific. https://doi.org/10.1142/9789812567222_0021

Deckert-Cruz, D. J., Tyler, R. H., Landmesser, J. E., & Rose, M. R. (1997). Allozymic differentiation in response to laboratory demographic selection of Drosophila melanogaster. *Evolution, 51*(3), 865—872. https://doi.org/10.1111/j.1558-5646.1997.tb03668.x

Fleming, J. E., Spicer, G. S., Garrison, R. C., & Rose, M. R. (1993). Two-dimensional protein electrophoretic analysis of postponed aging in Drosophila. *Genetica, 91*(1), 183—198. https://doi.org/10.1007/BF01435997

Graves, J. L., Luckinbill, L. S., & Nichols, A. (1988). Flight duration and wing beat frequency in long- and short-lived *Drosophila melanogaster*. *Journal of Insect Physiology, 34*(11), 1021—1026. https://doi.org/10.1016/0022-1910(88)90201-6

Luckinbill, L. S., Graves, J. L., Tomkiw, A., & Sowirka, O. (1988b). A qualitative analysis of some life-history correlates of longevity inDrosophila melanogaster. *Evolutionary Ecology, 2*(1), 85—94. https://doi.org/10.1007/BF02071591

Luckinbill, L. S., Grudzien, T. A., Rhine, S., & Weisman, G. (1989). The genetic basis of adaptation to selection for longevity in *Drosophila melanogaster*. *Evolutionary Ecology, 3*(1), 31—39. https://doi.org/10.1007/BF02147929

Maynard Smith, J., & Haigh, J. (1974). The hitch-hiking effect of a favourable gene. *Genetical Research, 23*(1), 23—35. https://doi.org/10.1017/S0016672300014634

Rose, M. R. (1984a). Genetic covariation in Drosophila life history: Untangling the data. *The American Naturalist, 123*(4), 565—569. https://doi.org/10.1086/284222

Rose, M. R. (1984b). Laboratory evolution of postponed senescence in *Drosophila melanogaster*. *Evolution, 38*(5), 1004—1010. https://doi.org/10.2307/2408434. JSTOR.

Service, P. M., & Rose, M. R. (1985). Genetic covariation among life-history components: The effect of novel environments. *Evolution, 39*(4), 943—945. https://doi.org/10.2307/2408694. JSTOR.

Service, P. M., Hutchinson, E. W., MacKinley, M. D., & Rose, M. R. (1985). Resistance to environmental stress in *Drosophila melanogaster* selected for postponed senescence. *Physiological Zoology, 58*(4), 380—389.

Tyler, R. H., Brar, H., Singh, M., Latorre, A., Graves, J. L., Mueller, L. D., Rose, M. R., & Ayala, F. J. (1993). The effect of superoxide dismutase alleles on aging in Drosophila. *Genetica, 91*(1), 143—149. https://doi.org/10.1007/BF01435994

CHAPTER THIRTY SIX

1988—89: The evolutionary logic of extending lifespan by dietary restriction

The standard paradigm

A classic result from the conventional gerontology literature is that rodent lifespan is greatly increased by caloric restriction, especially if such restriction begins before maturation. The landmark experiments (McCay et al., 1935, 1939, 1943) bred mice under normal conditions until the point of weaning, when a restricted diet would be introduced, with the result of delayed reproductive maturation. The historical term for this process was "retardation." After the period of dietary retardation was completed, the mice were reintroduced to standard caloric conditions and were able to begin reproduction. What made these experiments particularly intriguing was that it was then the only experimental treatment in mammals that increased longevity. It was also found in a variety of rodent stocks, even when the laboratory chow was varied. In addition, the health of calorie-restricted mice was generally similar to that of control populations, and sometimes better. Indeed, the metabolic rate of these mice was not reduced relative to the total mass of the given individuals, and some studies found that caloric restriction was associated with more functional patterns of protein synthesis (Birchenall-Sparks et al., 1985; Richardson et al., 1987).

To some, this phenomenon seemed too good to be true. This position was expressed by Masoro (1989), who questioned whether the ad libitum consumption of lab chow among the control rodents was itself pathological, with the improved longevity and health of the restriction groups simply a reflection of benefits from eating less of it. The possibility could be that the control food groups ate excessive calories, making the limited caloric intake more "natural" by comparison. Indeed, the "control" rodents in these experiments continued to gain weight after maturity, often interpreted as a tendency to obesity. This begs the question whether a "natural" diet could even be determined for such inbred experimental populations, long

Conceptual Breakthroughs in The Evolutionary Biology of Aging
ISBN: 978-0-12-821545-6
https://doi.org/10.1016/B978-0-12-821545-6.00041-8

maintained on laboratory chow that is usually based on cereal grains. As it stands, later-onset dietary restriction in rodents is one of the most common, though not universal, paradigms for prolonging longevity in laboratory mammals.

The conceptual breakthrough

Two competing hypotheses have been proposed to explain the phenomenon of increased longevity in response to caloric restriction in rodents. The first was proposed by Harrison and Archer (1988) and Holliday (1989), who thought that "life-stretching" during caloric restriction was an adaptive strategy for surviving intermittent periods of starvation. The second hypothesis came from Phelan and Austad (1989), who viewed life-stretching as a side effect of reduced nutrition in turn reducing reproduction, with reduced reproduction increasing longevity as a merely incidental side effect.

The viewpoint of Harrison and Archer (1988) was that life-stretching would be a necessity in populations of small mammals living in the wild, under conditions where resources were intermittently limited. The females that managed to survive these periods would be favored by natural selection, so long as they could reproduce when environmental conditions subsequently became more favorable. Holliday (1989) echoed this hypothesis by arguing that animals with higher caloric restriction tolerance would have greater Darwinian fitness due to their diversion of resources to the somatic body, and away from reproduction, in line with his disposable soma theory.

Phelan and Austad (1989) argued instead that natural selection would usually be too weak at later ages for deferred reproduction to make a significant contribution to net fitness. Instead, the trade-off between longevity and reproduction is one of immediate benefit, where the improved survival during periods of low-calorie nutrition is a side effect of the allocation of resources away from reproduction when such reproduction would in any case be impossible or minimal.

Impact: 4

This debate essentially hinged on whether the somatic benefits of a reduced caloric intake were part of an adaptive strategy or a merely incidental side effect of reducing the costs of reproduction. It should also be noted that there is a salient third position: that increased longevity with reduced caloric intake in the laboratory is an experimental artifact arising from preventing the obesity characteristic of control rodents, rodents that are typically kept in small enclosures with little to do but eat. These three hypotheses remain under debate to this day. From an evolutionary standpoint, choosing among these three hypotheses is complicated by their development from experimental research that typically employs inbred rodents consuming diets that they may not have been able to adapt to, due to their inbreeding.

References and further reading

Birchenall-Sparks, M. C., Roberts, M. S., Staecker, J., Hardwick, J. P., & Richardson, A. (1985). Effect of dietary restriction on liver protein synthesis in rats. *The Journal of Nutrition, 115*(7), 944—950. https://doi.org/10.1093/jn/115.7.944

Cerami, A. (1985). Hypothesis: Glucose as a mediator of aging. *Journal of the American Geriatrics Society, 33*(9), 626—634. https://doi.org/10.1111/j.1532-5415.1985.tb06319.x

Fernandes, G., Yunis, E. J., & Good, R. A. (1976). Influence of diet on survival of mice. *Proceedings of the National Academy of Sciences of the United States of America, 73*(4), 1279—1283.

Harrison, D. E., & Archer, J. R. (1988). Natural selection for extended longevity from food restriction. *Growth, Development, and Aging: GDA, 53*(1—2), 3.

Holliday, R. (1989). Food, reproduction, and Longevity: Is the extended lifespan of calorie-restricted animals an evolutionary adaptation? *BioEssays, 10*(4), 125—127. https://doi.org/10.1002/bies.950100408

Masoro, E. J., Katz, M. S., & McMahan, C. A. (1989). Evidence for the glycation hypothesis of aging from the food-restricted rodent model. *Journal of Gerontology, 44*(1), B20—B22. https://doi.org/10.1093/geronj/44.1.B20

McCay, C. M., Crowell, M. F., & Maynard, L. A. (1935). The effect of retarded growth upon the length of life span and upon the ultimate body size. *The Journal of Nutrition, 10*(1), 63—79. https://doi.org/10.1093/jn/10.1.63

McCay, C. M., Maynard, L. A., Sperling, G., & Barnes, L. L. (1939). Retarded growth, life span, ultimate body size and age changes in the albino rat after feeding diets restricted in calories. *The Journal of Nutrition, 18*(1), 1—13. https://doi.org/10.1093/jn/18.1.1

McCay, C. M., Sperling, G., & Barnes, L. L. (1943). Growth, ageing, chronic diseases, and life span in rats. *Archieves of Biochemistry, 2*, 469—479 (CABDirect).

Phelan, J. P., & Austad, S. (1989). Natural selection, dietary restriction, and extended longevity. *Growth, Development, and Aging: GDA, 53*, 4—6.

Richardson, A., Butler, J. A., Rutherford, M. S., Semsei, I., Gu, M. Z., Fernandes, G., & Chiang, W. H. (1987). Effect of age and dietary restriction on the expression of alpha 2u-globulin. *Journal of Biological Chemistry, 262*(26), 12821—12825. https://doi.org/10.1016/S0021-9258(18)45280-5

CHAPTER THIRTY SEVEN

1992: Selection for stress resistance increases lifespan

The standard paradigm

Preliminary work on the relationship between stress resistance and lifespan was performed by Service et al. (1985), who found that long-lived "O" populations produced using the method of Edney and Gill (1968) resisted several acute stresses across adulthood compared to their shorter-lived control "B" lines. The intuitively natural inference was that being better able to resist death due to any cause, either endogenous or external, would help these O populations live longer compared to the B populations that resembled their ancestors.

One commonly studied stress-response system in the Drosophila literature is the "heat-shock response," where flies subjected to high temperatures synthesize protein-chaperones to protect their other proteins from denaturing. Although the O's might have proven more resilient to this stressor, Service et al. (1985) found no such result. The explanation for this would not be resolved until the work of Kurapati et al. (2000), explained in Chapter 58.

Service et al. (1985) were successful, however, in demonstrating significant differences in the response to starvation and desiccation stressors. As expected, the long-lived O populations were more starvation and desiccation resistant than their B counterparts, in both males and females. Interestingly, metabolic rates were not observed to be different between the selection regimes. This finding was investigated more thoroughly in Service (1987), where it was found that O's stored greater deposits of lipids in their bodies relative to the B's and showed an age-specific increase in fat storage in female O's but not in males.

This work by Service et al. (1985, 1987) established an evolutionary relationship between longevity and stress resistance of some kind. But it did not establish a direct causal relationship between increased stress resistance and the postponement or slowing of aging in Drosophila. Some mediating factor could have been selectively favored by the O culture regime, which then

Conceptual Breakthroughs in The Evolutionary Biology of Aging
ISBN: 978-0-12-821545-6
https://doi.org/10.1016/B978-0-12-821545-6.00035-2

secondarily increased stress resistance. If a physiological mechanism underlying such postponed aging is indeed increased stress resistance, however, then selecting on stress resistance itself should increase lifespan.

The conceptual breakthrough

Rose et al. (1992) selected specifically for increased stress resistance in two experiments: one selected for resistance to total starvation; while the other selected for resistance to acute desiccation. In both cases, lifespan increased in parallel with the focally selected type of stress resistance. This supported the validity of the inferred causal importance of stress resistance for longevity. Subsequent experiments with yeast and nematodes (reviewed by Jazwinski, 1996), as discussed in Chapter 52, found that longer-lived mutants could be obtained by screening for resistance to acute stress, supporting the concept of a general association between lifespan and the capacity to resist acute stress.

Impact: 7

Rose et al. (1992) supplied additional experimental warrant for further research on the relationship between acute stress resistance and long-term adult survival, building on the work of Service et al. (1985) among others, research that continues to this day. However, later results would undermine the simplicity of the initial results obtained by Service et al. (1985) and Rose et al. (1992), as discussed in Chapter 50.

References and further reading

Edney, E. B., & Gill, R. W. (1968). Evolution of senescence and specific longevity. *Nature,* *220*(5164), 281—282. https://doi.org/10.1038/220281a0

Jazwinski, S. M. (1996). Longevity, genes, and aging. *Science, 273*(5271), 54—59. https://doi.org/10.1126/science.273.5271.54

Kurapati, R., Passananti, H. B., Rose, M. R., & Tower, J. (2000). Increased hsp22 RNA levels in Drosophila lines genetically selected for increased longevity. *The Journals of Gerontology: Series A, 55*(11), B552—B559. https://doi.org/10.1093/gerona/55.11.B552

Rose, M. R., Vu, L. N., Park, S. U., & Graves, J. L. (1992). Selection on stress resistance increases longevity in *Drosophila melanogaster. Experimental Gerontology, 27*(2), 241—250. https://doi.org/10.1016/0531-5565(92)90048-5

Service, P. M. (1987). Physiological mechanisms of increased stress resistance in *Drosophila melanogaster* selected for postponed senescence. *Physiological Zoology, 60*(3), 321—326.

Service, P. M., Hutchinson, E. W., MacKinley, M. D., & Rose, M. R. (1985). Resistance to environmental stress in *Drosophila melanogaster* selected for postponed senescence. *Physiological Zoology, 58*(4), 380—389.

CHAPTER THIRTY EIGHT

1992: In late adult life, mortality rates stop increasing

The standard paradigm

Given the enormous difficulties of studying age-specific mortality and age-specific fertility both experimentally and in the wild, *Homo sapiens* is the best characterized species demographically. Both evolutionary biologists and cytogerontologists were in agreement, prior to 1992, that in later adulthood aging continues relentlessly until all individuals are dead.

Many demographic models of aging used variants of the Gompertz (1825) equation ([Eq. 38.1] in its simplest form) to represent adult mortality rates as a function of age. Here, $\mu(x)$ denotes the mortality rate at age x, A is the age-independent mortality parameter, and α is the age-dependent parameter. The parameter α is often used to reflect the rate of aging, with mortality increasing progressively in an exponential fashion.

$$\mu(x) = Ae^{(\alpha x)} \tag{38.1}$$

This pattern of mortality rates was presumed to be a reasonable characterization of aging even under good conditions. In effect, it mathematically represented the near-universal assumption that progressive aging would result in a consistent decline in the likelihood of survival with age, a decline that would reach a point of guaranteed death in any finite cohort.

The nearly universal adoption of this model of exponentially rising death rates was first undermined by Greenwood and Irwin (1939). From a large body of human demographic data, Greenwood and Irwin (1939) noticed the occurrence of plateaus for mortality rates at advanced ages, ostensibly contrary to the Gompertz model and its variants. Unfortunately, their publication was generally neglected.

The conceptual breakthrough

In 1992, Carey et al. and Curtsinger et al. published two reports showing that, in large cohorts of medflies and fruit flies respectively, mortality rates stabilize late in adult life, rather than continuing to increase with age.

Conceptual Breakthroughs in The Evolutionary Biology of Aging
ISBN: 978-0-12-821545-6
https://doi.org/10.1016/B978-0-12-821545-6.00001-7

This finding was in direct conflict not only with the widespread use of Gompertzian models, but indeed in conflict with the near-universal assumption, common among every kind of biogerontologist, that there are finite upper limits to lifespan among aging species.

The study of Carey et al. (1992) used three different handling regimes with which to gather demographic data from adult medflies. The first two treatments consisted of total cohort sizes of 20,000 flies each. One of these treatments featured flies kept in cups containing flies of only one sex. The other treatment featured even lower numbers of flies maintained in cells. The third treatment regime was imposed on a massive 1.2 million medfly cohort, with numerous cages each started with 7200 flies, and subjected to higher crowding density and reproductive activity. Despite these different methods of handling, Carey et al. (1992) found that all three treatments exhibited mortality rates which plateaued for long periods after an initial period of demographic aging. In effect, demographic aging appeared to stop (Fig. 38.1).

Curtsinger's et al. (1992) observed a similar pattern in their analysis of an inbred *Drosophila melanogaster* cohort of 5751 male flies. Their mortality-rate estimates conformed to a Gompertz equation up to day 30, after which their observed mortality rates plateaued. However, presumably because of the much smaller cohort size in their study, the eventual mortality rate plateau that they observed was not as steady as those observed by Carey et al. (1992).

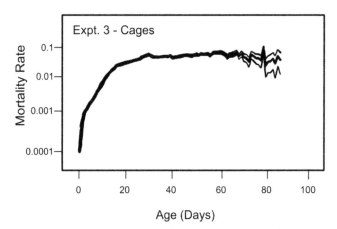

Age (Days)

Figure 38.1 Age-specific mortality in medfly cohorts. Trendline follows the results of experiment three, using 1,203,646 medflies in separate cages of 7,200, demonstrating a distinct mortality plateau in late-life ages. *Reproduced from data of Carey, J. R., Liedo, P., Orozco, D., & Vaupel, J. W. (1992). Slowing of mortality rates at older ages in large medfly cohorts.* Science, 258(5081), 457–461. (JSTOR).

Impact: 10

This was a revolutionary finding in biogerontology as a whole, and a direct challenge to the view that lifespan was inherently limited whenever the forces of natural selection decline with age according to the conventional interpretation of the evolutionary theory of aging (e.g., Rose, 1991). While there had been intimations of this finding before from human mortality data, especially in the research by Greenwood and Irwin (1939), the human case was regarded as too subject to differential treatment of the extremely old compared to the moderately old (e.g., Maynard Smith et al., 1999). But these two 1992 articles definitively shattered the view that demographic aging always continued unabated to extreme old ages. The value of these studies was strengthened by the varied handling protocols, the use of multiple replicate cohorts by Carey et al. (1992), the sheer size of the cohorts employed, and the use of two different species.

The long-term impact of this breakthrough research has been varied. Most biogerontologists have fastidiously ignored the salience of this research for their own work. By contrast, within the field of the evolutionary biology of aging these two publications have stimulated extensive theoretical and experimental research, much of which will be summarized below in subsequent chapters. The book *Does Aging Stop?* (Mueller et al., 2011) does an excellent job summarizing most of that work, albeit at a high level of difficulty.

References and further reading

Carey, J. R., Liedo, P., Orozco, D., & Vaupel, J. W. (1992). Slowing of mortality rates at older ages in large medfly cohorts. *Science, 258*(5081), 457–461 (JSTOR).

Curtsinger, J. W., Fukui, H. H., Townsend, D. R., & Vaupel, J. W. (1992). Demography of genotypes: Failure of the limited life-span paradigm in *Drosophila melanogaster*. *Science, 258*(5081), 461–463. https://doi.org/10.1126/science.1411541

Gompertz, B. (1825). On the nature of the function expressive of the law of human mortality, and on a new mode of determining the value of life contingencies. In F. R. S. c Esq (Ed.), *A letter to Francis Baily* (Vol. 115, pp. 513–583). Philosophical Transactions of the Royal Society of London. https://doi.org/10.1098/rstl.1825.0026, 1825.

Greenwood, M., & Irwin, J. O. (1939). The biostatistics of senility. *Human Biology, 11*(1), 1–23.

Kirkwood, T. B. L. (2015). Deciphering death: A commentary on Gompertz (1825). 'on the nature of the function expressive of the law of human mortality, and on a new mode of determining the value of life contingencies. *Philosophical Transactions of the Royal Society B: Biological Sciences, 370*(1666), 20140379. https://doi.org/10.1098/rstb.2014.0379

Maynard Smith, J., Barker, D., Finch, T., Kardia, S., Easton, B., Kirkwood, T., Le Grand, E., Nesse, R., Williams, C. G., & Partridge, L. (1999). The evolution of non-infectious and degenerative disease. In S. C. Stearns (Ed.), *Evolution in health and disease* (pp. 267–272). Oxford University Press. http://discovery.ucl.ac.uk/134752/.

Mueller, L. D., Rauser, C. L., & Rose, M. R. (2011). *Does aging stop?* Oxford: Oxford University Press.

Rose, M. R. (1991). *Evolutionary biology of aging*. Oxford University Press.

CHAPTER THIRTY NINE

1993—1995: Evolution of increased longevity among mammals, in the wild and the lab

The standard paradigm

Edney and Gill (1968) put forth the challenge that an important test of the evolutionary basis of senescence would be to compare populations undergoing different selection pressures arising from a simple difference in the timing of reproduction. By the early 1980s, this had effectively been accomplished (e.g., Luckinbill et al., 1984; Rose, 1984b; Rose & Charlesworth, 1980, 1981b; Sokal, 1970), but aside from invertebrates, there remained little evidence for this occurring in mammals.

The conceptual breakthrough

In 1993, Austad reported that insular possum populations which reproduced at later ages than their mainland counterparts had evolved greater longevities, in accordance with evolutionary expectations. Austad (1993) noted how island environments are typically "safer" environments, featuring a lack of apex predators, and ultimately more hospitable for longer-lived organisms. With a study focusing on two genetically distinct, radio-collared, Virginia opossum (*Didelphis virginiana)* populations, one on the mainland and the other from an island population, Austad (1993) probed whether the relative safety of each environment would impact longevity.

To score characteristics indicative of senescence, Austad (1993) tracked survivorship, litter sizes, and collagen degradation in tail tendon fibers. The results of his 1993 article supported Austad's intuition, as the insular opossum populations exhibited retarded senescence and reduced litter size, results which fall in line with the possibility of antagonistic pleiotropy between longevity and fecundity. In addition, the level of collagen degradation in mainland opossums was approximately double that of the insular population in later adult life.

Two years later in 1995, Nagai et al. reported that mouse lab populations that had been allowed to reproduce at later parental ages exclusively,

Conceptual Breakthroughs in The Evolutionary Biology of Aging
ISBN: 978-0-12-821545-6
https://doi.org/10.1016/B978-0-12-821545-6.00044-3

likewise had evolved increased lifespans. The experimental set up consisted of three lines "121," "141," and "221." Selection was imposed on 121 and 141, relative to the control 221 line, by delaying the age at first reproduction. Nagai et al. (1995) concluded that selection for delayed reproduction had resulted in the evolution of improved total lifespan, which matched the expectations of the evolutionary theory of aging.

Impact: 6

The findings of Austad (1993) and Nagai et al. (1995) underscored the generality of the evolutionary explanation for the existence and timing of aging. While many laboratories had shown that aging would readily evolve in insect populations in response to changes in the timing of reproduction, up until 1993 there was little direct evidence that the evolution of aging would proceed in a similar manner among mammalian species. Research on the evolution of aging with mammalian species continues (McHugh & Burke, 2022), particularly rabbits (Theilgaard et al., 2007).

References and further reading

Austad, S. N. (1993). Retarded senescence in an insular population of Virginia opossums (*Didelphis virginiana*). *Journal of Zoology, 229*(4), 695–708. https://doi.org/10.1111/j.1469-7998.1993.tb02665.x

Edney, E. B., & Gill, R. W. (1968). Evolution of senescence and specific longevity. *Nature, 220*(5164), 281–282. https://doi.org/10.1038/220281a0

Luckinbill, L. S., Arking, R., Clare, M. J., Cirocco, W. C., & Buck, S. A. (1984). Selection for delayed senescence in *Drosophila melanogaster*. *Evolution, 38*(5), 996–1003. https://doi.org/10.2307/2408433. JSTOR.

McHugh, K. M., & Burke, M. K. (2022). From microbes to mammals: The experimental evolution of aging and longevity across species. *Evolution, 76*(4), 692–707. https://doi.org/10.1111/evo.14442

Mertz, D. B. (1975). Senescent decline in flour beetle strains selected for early adult fitness. *Physiological Zoology, 48*(1), 1–23. https://doi.org/10.1086/physzool.48.1.30155634

Nagai, J., Lin, C. Y., & Sabour, M. P. (1995). Lines of mice selected for reproductive longevity. *Growth, Development, and Aging: GDA, 59*(3), 79–91.

Rose, M. R. (1984a). Genetic covariation in Drosophila life history: Untangling the data. *The American Naturalist, 123*(4), 565–569. https://doi.org/10.1086/284222

Rose, M. R. (1984b). Laboratory evolution of postponed senescence in *Drosophila melanogaster*. *Evolution, 38*(5), 1004–1010. https://doi.org/10.2307/2408434. JSTOR.

Rose, M., & Charlesworth, B. (1980). A test of evolutionary theories of senescence. *Nature, 287*(5778), 141–142. https://doi.org/10.1038/287141a0

Rose, M. R., & Charlesworth, B. (1981b). Genetics of life history in *Drosophila melanogaster*. II. Exploratory selection experiments. *Genetics, 97*(1), 187–196.

Sokal, R. R. (1970). Senescence and genetic load: Evidence from Tribolium. *Science (New York, N.Y.), 167*(3926), 1733–1734. https://doi.org/10.1126/science.167.3926.1733

Theilgaard, P., Sánchez, J. P., Pascual, J. J., Berg, P., Friggens, N. C., & Baselga, M. (2007). Late reproductive senescence in a rabbit line hyper selected for reproductive longevity, and its association with body reserves. *Genetics Selection Evolution, 39*(2), 207. https://doi.org/10.1186/1297-9686-39-2-207

CHAPTER FORTY

1993: Evolutionary physiology of dietary restriction

The standard paradigm

In reductionist cytogerontology, the benefits of caloric restriction for longevity are usually attributed to such molecular phenomena as reduced free-radical production and the like. But the analysis of Phelan and Austad (1989) suggests that reduced costs of reproduction resulting from whole-organism physiological mechanisms are more likely to underlie the benefits of caloric restriction, as detailed in Chapter 36.

An early commentary on life-history evolution from George C. Williams (1966) emphasized the inevitable trade-off of resources between reproduction and longevity, especially the resource of caloric energy. Any calories devoted to the production of gametes come at the potential expense of converting said energy into fat, where greater fat reserves improve the likelihood of surviving conditions of starvation. Service and Rose (1985) and Service (1987) provided early evidence that fat storage and starvation resistance were strong indicators for longevity, an inference supported by the results of Rose et al. (1992) described in Chapter 37. But less was known about the relationship of this fat diversion to reproduction.

The conceptual breakthrough

Chippindale et al. (1993) studied dietary restriction in Drosophila populations that had evolved different patterns of aging, using tools and ideas from evolutionary physiology. They found that dietary restriction caused a substantial drop in female reproduction, along with a considerable increase in starvation resistance. Given the positive relationship between the evolution of enhanced stress resistance and longevity in the evolution of these Drosophila, as well as the evidence for antagonistic pleiotropy between early reproduction and longevity in the same populations, they were motivated to test whether the longevity benefits of dietary restriction were at least partly mediated by the same physiological machinery as that which underpinned the evolution of aging.

Conceptual Breakthroughs in The Evolutionary Biology of Aging
ISBN: 978-0-12-821545-6
https://doi.org/10.1016/B978-0-12-821545-6.00005-4

To accomplish this, they compared the five Methuselah "O" populations to the five control "B" populations of Rose (1984b), with two separate food regimens: either large or small amounts of yeast paste spread over agar food plates. The level of nutrient-rich yeast paste was expected to induce either mating or fat storage, depending on the treatment group. Additionally, Chippindale et al. (1993) also switched the diet from high yeast to low yeast or low yeast to high yeast for a subset of the treatments.

Their findings revealed a striking pattern of adult flies switching from high reproductive output with lowered fat reserves to low reproductive output with increased fat reserves, as diet quality switched from good to poor. This switch took about 3 days. Rather than a slow playing out of different patterns of molecular damage, this pattern depended on organ-level allocation of energetic resources between reproductive structures and fat reserves. Thus, this is another case where the physiology of aging is interpretable on evolutionary principles, as is common in evolutionary physiology generally, rather than the prevalent cell-molecular notion of cumulative damage.

Impact: 7

Chippindale et al. (1993) provided important empirical evidence concerning the mechanistic basis for the reallocation of resources between survival and reproduction in adult *Drosophila melanogaster*, an allocation that is fully reversible. Such Drosophila evolutionary physiology research would be developed on a large scale, following the initial leads of Service et al. (1985) and Chippindale et al. (1993), much of this work collected in the book *Methuselah Flies* (Rose et al., 2004).

Broadly speaking, the intercalation of these themes has become common across a variety of studies in the field of evolutionary physiology. It is unclear whether this physiological research on the experimental evolution of aging has been particular influential. The thinking of George C. Williams is probably the most important influence on this field, as transmitted by his 1966 article, rather than his 1957 article on the evolution of aging. In a sense, the field of evolutionary physiology evolved in large part from the ideas of evolutionary ecology, a subfield that connects evolutionary biology and ecology, especially the theme of trade-offs that Williams (1957, 1966) long emphasized (See also the Mueller [2020] book in the Conceptual Breakthroughs series).

References and further reading

Chippindale, A. K., Leroi, A. M., Kim, S. B., & Rose, M. R. (1993). Phenotypic plasticity and selection in Drosophila life-history evolution. I. Nutrition and the cost of reproduction. *Journal of Evolutionary Biology, 6*(2), 171–193. https://doi.org/10.1046/j.1420-9101.1993.6020171.x

Mueller, L. D. (2020). *Conceptual breakthroughs in evolutionary ecology.* Academic Press, an imprint of Elsevier.

Phelan, J. P., & Austad, S. (1989). Natural selection, dietary restriction, and extended longevity. *Growth, Development, and Aging: GDA, 53,* 4–6.

Rose, M. R. (1984b). Laboratory evolution of postponed senescence in *Drosophila melanogaster. Evolution, 38*(5), 1004–1010. https://doi.org/10.2307/2408434. JSTOR.

Rose, M. R., Passananti, H. B., & Matos, M. (2004). *Methuselah flies: A case study in the evoultion of aging.* World Scientific Pub.

Rose, M. R., Vu, L. N., Park, S. U., & Graves, J. L. (1992). Selection on stress resistance increases longevity in *Drosophila melanogaster. Experimental Gerontology, 27*(2), 241–250. https://doi.org/10.1016/0531-5565(92)90048-5

Service, P. M. (1987). Physiological mechanisms of increased stress resistance in *Drosophila melanogaster* selected for postponed senescence. *Physiological Zoology, 60*(3), 321–326.

Service, P. M., Hutchinson, E. W., MacKinley, M. D., & Rose, M. R. (1985). Resistance to environmental stress in *Drosophila melanogaster* selected for postponed senescence. *Physiological Zoology, 58*(4), 380–389.

Service, P. M., & Rose, M. R. (1985). Genetic covariation among life-history components: The effect of novel environments. *Evolution, 39*(4), 943–945. https://doi.org/10.2307/2408694. JSTOR.

Williams, G. C. (1957). Pleiotropy, natural selection, and the evolution of senescence. *Evolution, 11*(4), 398–411. https://doi.org/10.2307/2406060. JSTOR.

Williams, G. C. (1966). Natural selection, the costs of reproduction, and a refinement of lack's principle. *The American Naturalist, 100*(916), 687–690.

CHAPTER FORTY ONE

1993: Genetic association between dauer metabolic arrest and increased lifespan

The standard paradigm

The classic work of Riddle et al. (1981) with *C. elegans* pioneered the search for mutants that were not only stress resistant but also long-lived. A primary focus of these studies was on the genetic predisposition in nematodes to divert development toward formation of dauer larvae. Dauers are an alternative third larval stage at which development may arrest, typically induced by crowding or limited nutrients. Dauer larvae appear to be exempt from normal aging in that they can resume development upon restoration of a benign environment after 1—2 months, then yielding an adult of normal lifespan. *C. elegans* nematodes thus halt their progression to adulthood and slow much of their metabolism, under conditions of nutritional stress. For mutations to genes comprising the dauer-formation pathway, the abbreviation *daf* ("dauer larva formation") is used. Riddle et al. (1981) explored many dauer-constitutive mutants (e.g., from *daf-2, daf-4,* and *daf-7* genes), but at this time, it was not clear what the physiological action of the first nematode "longevity" mutants was, nor was it initially possible to clone their genes.

The conceptual breakthrough

The molecular genetics of the regulation of lifespan in *C. elegans* was greatly clarified by the discovery by Kenyon et al. (1993) that *daf* mutations also strongly impact *adult* longevity. Two genes involved were named *daf-2* and *daf-16*, which lay within a single branch of a bifurcated dauer-formation pathway. Kenyon et al. (1993) discovered that *C. elegans* hermaphrodites (which comprise 99.8% of most *C. elegans* populations), when they carry one of three mutant alleles of the *daf-2* gene, have a longevity twice that of their wild-type counterparts. But these mutant nematodes revert to normal lifespan if they receive a loss-of-function mutation to the *daf-16* gene. This gene—gene interaction provided evidence of epistasis, i.e.,

Conceptual Breakthroughs in The Evolutionary Biology of Aging
ISBN: 978-0-12-821545-6
https://doi.org/10.1016/B978-0-12-821545-6.00018-2

sequential action in a shared genetic pathway. Kenyon et al. (1993) were able to take advantage of the temperature-sensitivity (*ts*) characteristic of *daf-2* mutations, which are nonpenetrant at temperatures up to 15°C, to avoid developmental arrest as dauer larvae and thus to observe the otherwise-covert trait of life extension after shift of nearly-mature larvae to nonpermissive temperatures. Thus the long adult survival of these mutants did not arise merely from arrested growth, but rather was produced by altered downstream genetic regulation. An appealing interpretation is that such effects had evolved to enable dauer larvae to survive extended periods of food deprivation or harsh environments.

Kenyon et al. (1993) compared *daf-2(e1370)* mutant worms with wild-type *C. elegans* under various temperature regimens and with or without laser ablation of germline cells in hermaphrodites. Their aim was to determine how sensitive *daf* mutant life extension effects were to temperature manipulation, and to determine if reproduction lowers life expectancy as found in earlier studies (Van Voorhies, 1992). The extended life span trait was conserved in *daf-2(e1370)* at several temperatures, while normal motility and feeding peristalsis were unimpaired. After ablating the precursors of the germline of the hermaphroditic *C. elegans*, Kenyon et al. (1993) found no significant alterations in lifespan, indicating that the production of germ cells themselves was not directly limiting the wild-type lifespan.

Over the next few decades, multiple *C. elegans* loci were identified that had alleles which prolonged lifespan and were also involved in nematode dauer formation, e.g. (Larsen et al., 1995). Kenyon et al. (1993) noted the similarities of the *daf-2* mutants to Drosophila lines with extended life span (Rose, 1984b), where each demonstrated increased resistance to starvation and desiccation. They then proposed the existence of a conserved regulatory mechanism shared between the two species. Somewhat analogously to the work combining evolutionary physiology with experimental evolution in Drosophila, nematode lifespan regulation was mechanistically tied to their responses to nutrient restriction in these mutants. This was the start of a period of extensive research on dauer loci as sources of mutants which could greatly extend nematode longevity.

Impact: 8

C. elegans is the chief invertebrate alternative to Drosophila as a biomedical model system for aging research. Its strength is that it does not suffer the inbreeding depression that afflicts inbred, and thus most mutant, strains of Drosophila, or indeed the rodent stocks that are used in almost all mammalian aging research. This is because *C. elegans* reproduces primarily through hermaphroditic self-fertilization, so that alleles are selected according to their effects when homozygous. Thus its evolutionary genetics are radically unlike those of outbreeding diploid populations, which make up the vast majority of animal species, including the medically pertinent human species.

C. elegans aging research had exceptional power in the pregenomic era, because it could more reliably dissect the effects of single genes of major effect on aging and its cognate characters. Furthermore, species with reproductive systems like *C. elegans* primarily evolve through the substitution of single mutations of large beneficial effects, providing a strong connection between laboratory genetic manipulation and the possibilities for *C. elegans* evolution, with respect to aging and any other functional character.

Yet these same strengths as an experimental system make *C. elegans* a notably poor choice for *modeling* the evolution and genetics of aging in species that are primarily outbred and diploid. In contrast, Drosophila systems aptly serve this biomedical function, as discussed further in Chapter 57.

References and further reading

Kenyon, C., Chang, J., Gensch, E., Rudner, A., & Tabtiang, R. (1993). A *C. elegans* mutant that lives twice as long as wild type. *Nature, 366*(6454), 461–464. https://doi.org/10.1038/366461a0

Larsen, P. L., Albert, P. S., & Riddle, D. L. (1995). Genes that regulate both development and longevity in *Caenorhabditis elegans. Genetics, 139*(4), 1567–1583. https://doi.org/10.1093/genetics/139.4.1567

Riddle, D. L., Swanson, M. M., & Albert, P. S. (1981). Interacting genes in nematode dauer larva formation. *Nature, 290*(5808), 668–671. https://doi.org/10.1038/290668a0

Rose, M. R. (1984b). Laboratory evolution of postponed senescence in *Drosophila melanogaster. Evolution, 38*(5), 1004–1010. https://doi.org/10.2307/2408434. JSTOR.

Van Voorhies, W. A. (1992). Production of sperm reduces nematode lifespan. *Nature, 360*(6403), 456–458. https://doi.org/10.1038/360456a0

CHAPTER FORTY TWO

1992–95: Experimental evolution of aging is connected to development

The standard paradigm

Lints in 1978 and 1988 provided an explicit developmental theory of aging, in which differentiation during growth is responsible for later senescence. Theories of this kind hold that development is a continuous process that continues into adult life, with senescence a major feature of the later parts of the overall process (In some respects, this research builds on some of Bidder's [1932] thinking, which we covered in Chapter 11). Aging, and its impacts on length of lifespan, are seen by Lints as the result of a lifelong series of epigenetic events shaping differentiation and aging, all of it underpinned by specific regulatory pathways. This view of aging ties development and aging closely as functions of the same processes, with presumably the same sets of genes acting in an age-specific manner.

The chief finding supporting this view was experimental correlations between delayed development and longevity. The best evidence for such developmental correlations with aging was the association between delayed development and increased lifespan in calorically restricted rodents (McCay et al., 1935, 1939, 1943). In the context of natural history, there is also a general correspondence between the length of development and maximum longevity among species observed in comparative studies (vid. Comfort, 1979; Rose, 1991).

The conceptual breakthrough

Given the natural history and nutritional experiments that supported the developmental theory of aging, as well as the mutant *C. elegans* research that genetically connected dauer formation to longevity (vid. Chapter 41), evolutionary research on aging needed to integrate and explain the relationship between development and aging. Accordingly, Chippindale et al. (1994) compared the duration of development among Drosophila

Conceptual Breakthroughs in The Evolutionary Biology of Aging
ISBN: 978-0-12-821545-6
https://doi.org/10.1016/B978-0-12-821545-6.00034-0

populations that had experimentally evolved different lifespans. In a comparison of the longer-lived O populations with the shorter-lived B populations, it was found that selection regimes which delayed first ages of reproduction were associated with prolonged development. With the prolongation of development, however, came reduced early fecundity. This relationship evidently is a manifestation of antagonistic pleiotropy.

This work of Chippendale et al. (1994) differed somewhat from that of Partridge and Fowler (1992) and Zwaan (1995), who found developmental time differences that may have arisen from uncontrolled larval densities or inbreeding depression in the stocks that they used. Some results were consistent among all three studies, such as the general trend of longer-lived populations demonstrating slower developmental time.

Subsequent work selected directly for more rapid development in Drosophila. These rapidly developing populations had shorter lifespans, smaller bodies, and reduced total reproductive output (Chippindale et al., 1997, 2004), which corroborated the findings of Chippendale et al. (1994). Altogether, this body of research revealed that the connections between development and aging were underpinned by the forces of natural selection acting in functional genetic contexts that featured antagonistic pleiotropy.

Impact: 8

Within the context of the evolutionary biology of aging, the connection between developmental timing and aging is generally seen and experimentally analyzed in terms of the timing of patterns of natural selection, with selection tuning both the conclusion of development and the start of aging. As both development and aging are now easily and commonly manipulated using the tools of experimental evolution, the relationship between development and aging is no longer a matter of conjecture about general features of ontogeny stretching from the beginning to the end of life. Thus what began as confused conceptual conjectures became well-established correlations arising from patterns of selection and pleiotropy. The evolutionary genetics of development do interact with the evolution of aging, particularly in ways that can be interpreted using the experimental and analytical tools of the evolutionary biology of aging.

References and further reading

Bidder, G. P. (1932). Senescence. *BMJ*, *2*(3742), 583—585. https://doi.org/10.1136/bmj.2.3742.583

Chippindale, A. K., Alipaz, J. A., Chen, H.-W., & Rose, M. R. (1997). Experimental evolution of accelerated development in Drosophila. 1. Developmental speed and larval survial. *Evolution*, *51*(5), 1536—1551. https://doi.org/10.1111/j.1558-5646.1997.tb01477.x

Chippindale, A. K., Alipaz, J. A., & Rose, M. R. (2004). Experimental evolution of accelerated development in Drosophila. 2. Adult fitness and the fast development syndrome. *Methuselah Flies*, (2004), 413—435. https://doi.org/10.1142/9789812567222_0034

Chippindale, A. K., Hoang, D. T., Service, P. M., & Rose, M. R. (1994). The evolution of development in Drosophila melanogaster selected for postponed senescence. *Evolution*, *48*(6), 1880—1899. https://doi.org/10.1111/j.1558-5646.1994.tb02221.x

Comfort, A. (1979). *The biology of senescence* (3rd ed.). Edinburgh and London: Churchill Livingstone.

Lints, F. A. (1978). *Genetics and ageing interdisciplinary topics in gerontology*. Karger basil.

Lints, F. A., & Lints, C. V. (1969). Influence of preimaginal environment on fecundity and ageing in *Drosophila melanogaster* hybrids I. Preimaginal population density. *Experimental Gerontology*, *4*(4), 231—244. https://doi.org/10.1016/0531-5565(69)90011-4

Lints, F. A., & Soliman, M. H. (Eds.). (1988). *Drosophila as a model organism for ageing studies* (pp. 98—119). Springer US. https://doi.org/10.1007/978-1-4899-2683-8

McCay, C. M., Crowell, M. F., & Maynard, L. A. (1935). The effect of retarded growth upon the length of life span and upon the ultimate body size. *The Journal of Nutrition*, *10*(1), 63—79. https://doi.org/10.1093/jn/10.1.63

McCay, C. M., Maynard, L. A., Sperling, G., & Barnes, L. L. (1939). Retarded growth, life span, ultimate body size and age changes in the albino rat after feeding diets restricted in calories. *The Journal of Nutrition*, *18*(1), 1—13. https://doi.org/10.1093/jn/18.1.1

McCay, C. M., Sperling, G., & Barnes, L. L. (1943). Growth, ageing, chronic diseases, and life span in rats. *Achieves of Biochemistry*, *2*, 469—479 (CABDirect).

Partridge, L., & Fowler, K. (1992). Direct and correlated responses to selection on age at reproduction in *Drosophila melanogaster*. *Evolution*, *46*(1), 76—91. https://doi.org/10.1111/j.1558-5646.1992.tb01986.x

Rose, M. R. (1991). *Evolutionary biology of aging*. Oxford University Press.

Zwaan, B., Bijlsma, R., & Hoekstra, R. F. (1995). Artificial selection for developmental time in *Drosophila melanogaster* in relation to the evolution of aging: Direct and correlated responses. *Evolution*, *49*(4), 635—648. https://doi.org/10.1111/j.1558-5646.1995.tb02300.x

CHAPTER FORTY THREE

1994—96: Evidence for mutation accumulation affecting virility and aging

The standard paradigm

Up until the mid-1990s, most evolutionary-genetic evidence supported the involvement of antagonistic pleiotropy in the evolution of aging, especially genetic trade-offs between early reproduction and later adult survival in the experimental evolution of accelerated or delayed aging. Less attention was directed to the other possible mechanism for an evolutionary explanation for aging, mutation accumulation. The hypothesis first outlined by Peter Medawar in 1952 proposes that as the force of natural selection becomes weaker with each successive age class, the buildup of mutations of deleterious effects becomes more pronounced. As the age of onset increases for these allelic effects, natural selection is unable to filter and prevent this accumulation.

Promislow and Tatar (1998) outlined four possible approaches to testing for mutation accumulation. They proposed that the chief tests would involve comparing the age-specific fitness characters between populations, reversing selection for long-lived populations while measuring early fitness to see if the patterns are better explained by antagonistic pleiotropy, P-element mutations, and the effects of inbreeding depression in inbred lines.

The conceptual breakthrough

Hughes and Charlesworth (1994) as well as Hughes (1995a) produced evidence supporting the action of mutation accumulation in the aging of virility in Drosophila. Their studies employed 17,529 males constructed using third chromosomes extracted using balancer chromosomes from a recombinant inbred stock derived ultimately from the Ives population of *Drosophila melanogaster* studied by Rose and Charlesworth (e.g., 1981). Hughes (1995a) measured additive genetic covariances for fitness-related characters like age-specific survivorship, age-specific competitive ability, body mass, and fertility.

Conceptual Breakthroughs in The Evolutionary Biology of Aging
ISBN: 978-0-12-821545-6
https://doi.org/10.1016/B978-0-12-821545-6.00036-4

They found that the genetic variability among lines increased with adult age, in keeping with the mutation accumulation mechanism for the evolutionary genetics of aging. Hughes (1995b) as well as Charlesworth and Hughes (1996) looked at inbreeding depression effects, which are predicted to be greater with adult age when mutation accumulation is the predominant mechanism for the evolution of aging. Specifically, they studied the aforementioned third-chromosome lines that differed according to whether that chromosome was homozygous or heterozygous. They assayed both male survival and male mating success. Their findings clearly support the action of mutation accumulation, as opposed to antagonistic pleiotropy.

Impact: 8

Borash et al. (2007) later found corroborative results in 10 related Drosophila populations, using different methods. Exploring many of the same fitness characters as Hughes (1995a), Borash et al. (2007) used outbred populations derived from the IV's, the longer-lived O populations, and the shorter-lived B populations of Rose (1984), comparing them for male reproductive success, female fecundity, and overall longevity. In addition, hybrids were created from O females with B males and vice versa to track their mating success. The goal of this study was to determine whether the B populations were affected by mutation accumulation. Borash et al. (2007) addressed the question of whether selection for late-life fertility would purge alleles that produce a decline in later fitness components, presumably due to mutation accumulation in the Bs. The findings of Borash et al. (2007) on male reproduction were in line with expectations, where an absence of selection on the B populations at later ages resulted in decreased virility with directional dominance in favor of virility, suggesting the evolutionary accumulation of recessive deleterious alleles.

The studies of Kosuda (1985), Mueller (1987), Hughes (1995a), and Borash et al. (2007) are alike in that they feature lines or genetic constructs in which the action of selection at later ages has been impeded by early breeding of experimental stocks. In addition, the protocols used by Kosuda (1985), Mueller (1987), and Hughes (1995a) featured periods of possible inbreeding during which mutation accumulation is expected to be accelerated. Overall, this body of work supports the possibility that mutation accumulation can be an important genetic mechanism for the evolution of aging when populations have few to no opportunities for reproduction later in life, especially if this occurs in conjunction with inbreeding.

By the 1990s, it was apparent that there was credible evidence for the involvement of both conceivable genetic mechanisms for the evolution of aging: antagonistic pleiotropy and mutation accumulation. Later work in the genomic era would reveal the relative importance of these two alternative genetic mechanisms.

References and further reading

Borash, D. J., Rose, M. R., & Mueller, L. D. (2007). Mutation accumulation affects male virility in Drosophila selected for later reproduction. *Physiological and Biochemical Zoology, 80*(5), 461–472. https://doi.org/10.1086/520127

Charlesworth, B. (2001). Patterns of age-specific means and genetic variances of mortality rates predicted by the mutation-accumulation theory of ageing. *Journal of Theoretical Biology, 210*(1), 47–65. https://doi.org/10.1006/jtbi.2001.2296

Charlesworth, B., & Hughes, K. A. (1996). Age-specific inbreeding depression and components of genetic variance in relation to the evolution of senescence. *Proceedings of the National Academy of Sciences of the United States of America, 93*(12), 6140–6145.

Hughes, K. A. (1995a). The evolutionary genetics of male life-history characters in Drosophila melanogaster. *Evolution, 49*(3), 521–537. https://doi.org/10.1111/j.1558-5646.1995.tb02284.x

Hughes, K. A. (1995b). The inbreeding decline and average dominance of genes affecting male life-history characters in *Drosophila melanogaster. Genetical Research, 65*(1), 41–52. https://doi.org/10.1017/S0016672300032997

Hughes, K. A., & Charlesworth, B. (1994). A genetic analysis of senescence in Drosophila. *Nature, 367*(6458), 64–66. https://doi.org/10.1038/367064a0

Kosuda, K. (1985). The aging effect on male mating activity in *Drosophila melanogaster. Behavior Genetics, 15*(3), 297–303. https://doi.org/10.1007/BF01065984

Medawar, P. B. (1952). *An unsolved problem of biology.* H.K. Lewis and Co.

Mueller, L. (1987). Evolution of accelerated senescence in laboratory populations of Drosophila. *Proceedings of the National Academy of Sciences of the United States of America, 84,* 1974–1977. https://doi.org/10.1073/pnas.84.7.1974

Promislow, D. E., & Tatar, M. (1998). Mutation and senescence: Where genetics and demography meet. *Genetica, 102–103*(1–6), 299–314.

Rose, M. R. (1984). Laboratory evolution of postponed senescence in *Drosophila melanogaster. Evolution, 38*(5), 1004–1010. https://doi.org/10.2307/2408434. JSTOR.

Rose, M. R., & Charlesworth, B. (1981). Genetics of life history in Drosophila melanogaster. I. Sib analysis of adult females. *Genetics, 97*(1), 173–186.

CHAPTER FORTY FOUR

1996—98: Physiological research on evolution of aging supports organismal mechanisms

The standard paradigm

The reductionist cell-molecular paradigm for aging assumes that the key controls of aging are to be found in such phenomena as DNA repair, translation-error catastrophe, and free-radical damage. Molecular theories of this kind rely on either endogenous failure of essential cell functions or exogenous agents of damage that disrupt or overwhelm somatic cells. A major challenge these proposed mechanisms face is that it is difficult to demonstrate that cellular or molecular damage is the cause of aging directly. On the other hand, this gives them reduced falsifiability, making it very difficult to collect evidence eliminating such hypotheses from contention. The destruction of somatic mutation and translation-error catastrophe theories in the 1960s and 1970s were notable exceptions, in that multiple scientists went to considerable trouble to falsify them (see Chapters 13 and 14). Fortunately for cell-molecular biologists, major aging research agencies like the U.S. National Institute of Aging no longer fund research of that kind, leaving the standard paradigm of cell-molecular theories of aging as intact as the geocentric theory of astrophysics was during the lives of Giordano Bruno and Galileo.

The evolutionary biology of aging starts from entirely different premises than those of cell-molecular reductionist theories of aging. The challenge that the evolutionary biology of aging faces is the difficulty of proceeding from patterns of natural selection through genetic machinery to the specific physiological mechanisms involved in the loss of function at later adult ages. As has been shown already in Chapter 29, early efforts in that direction included the work of Service et al. (1985) and Graves et al. (1988), who studied the organismal physiology of Drosophila populations that had evolved greater lifespans due to shifts in Hamilton's forces of natural selection.

Conceptual Breakthroughs in The Evolutionary Biology of Aging
ISBN: 978-0-12-821545-6
https://doi.org/10.1016/B978-0-12-821545-6.00004-2

The conceptual breakthrough

In a series of publications from 1996 to 1998, physiologists Djawdan, Bradley, Gibbs, and others probed the biochemical mechanisms underlying the increased stress resistance previously shown to underlie the experimental evolution of increased lifespan by Service et al. (1985), Service (1987), and Rose et al. (1992). They found that increases in physiological stress resistance were determined entirely by whole-organism storage of reserve substances and their rates of loss, not widely supposed cell-molecular mechanisms like somatic mutation. By contrast, repeated attempts to find associations between experimentally evolved increased lifespan and commonly hypothesized cell-molecular mechanisms of aging failed, though such failures were not published.

Djawdan et al. (1996) examined the trade-off that occurs between reproduction and longevity in the experimental evolution of aging by measuring the metabolic energy allocation of five long-lived *Drosophila melanogaster* populations relative to controls with shorter lifespans. Tabulating the relative storage of carbohydrates and fats in the Drosophila body relative to the energy devoted to fecundity, Djawdan et al. (1996) found that the total energy (J) present in both short and long-lived flies was similar. The main difference between them was that long-lived flies stored more lipid and carbohydrate, while short-lived populations expended more calories on increased early fecundity. As the first study to measure allocations of metabolic energy to both storage and reproduction, Djawdan et al. (1996) marks a significant step in the elucidation of the physiological basis of antagonistic pleiotropy. Later work by Djawdan et al. (1998) would demonstrate that caloric stores also were a good indicator of death prediction from starvation.

The evolution of increased desiccation resistance in experimentally evolved longer-lived populations was found to have different mechanistic foundations from those of increased starvation resistance (Gibbs et al., 1997; Chippindale et al., 1998). It was found that the accumulation of calories was not a good indicator for survival under conditions of acute desiccation, as both long and shorter-lived flies demonstrated similar metabolic rates under desiccating conditions. This finding falsified the hypothesis that caloric stores would be responsible for desiccation resistance, as proposed by Hoffman and Parsons (1994). Gibbs et al. (1997) and Chippindale et al. (1998) instead found that increased desiccation resistance was dependent on increased water retention and reduction of transcuticular water loss.

Impact: 7

Cell-molecular theories of aging, which rely heavily on the premise that aging is a merely physiological process, rarely survive strong-inference tests. The evolutionary biology of aging is based on theories of aging that not only survive strong-inference tests, but also allow experimental evolutionists to create populations that can be used to pursue physiological hypotheses for aging. The study of the evolutionary physiology of aging started by Service et al. (1985) is a line of investigation that continues to this day, while Djawdan et al. (as described above) took it to a more mechanistic level in their work on biochemical storage and metabolic utilization of such stores. As will be shown in subsequent chapters, the molecular genetics underlying such biochemical differentiation would be revealed by genomic and transcriptomic work on experimentally evolved populations.

References and further reading

Chippindale, A. K., Gibbs, A. G., Sheik, M., Yee, K. J., Djawdan, M., Bradley, T. J., & Rose, M. R. (1998). Resource acquisition and the evolution of stress resistance in *Drosophila melanogaster*. *Evolution, 52*(5), 1342—1352. https://doi.org/10.1111/j.1558-5646.1998.tb02016.x

Djawdan, M., Chippindale, A. K., Rose, M. R., & Bradley, T. J. (1998). Metabolic reserves and evolved stress resistance in *Drosophila melanogaster*. *Physiological Zoology, 71*(5), 584—594. https://doi.org/10.1086/515963

Djawdan, M., Sugiyama, T. T., Schlaeger, L. K., Bradley, T. J., & Rose, M. R. (1996). Metabolic aspects of the trade-off between fecundity and longevity in *Drosophila melanogaster*. *Physiological Zoology, 69*(5), 1176—1195.

Gibbs, A. G., Chippindale, A. K., & Rose, M. R. (1997). Physiological mechanisms of evolved desiccation resistance in *Drosophila melanogaster*. *The Journal of Experimental Biology, 200*(Pt 12), 1821—1832.

Graves, J. L., Luckinbill, L. S., & Nichols, A. (1988). Flight duration and wing beat frequency in long- and short-lived *Drosophila melanogaster*. *Journal of Insect Physiology, 34*(11), 1021—1026. https://doi.org/10.1016/0022-1910(88)90201-6

Hoffmann, A. A., & Parsons, P. A. (1994). *Evolutionary genetics and environmental stress (Reprint)*. Oxford University Press, ISBN 0-19-854081-7.

Rose, M. R., Vu, L. N., Park, S. U., & Graves, J. L. (1992). Selection on stress resistance increases longevity in *Drosophila melanogaster*. *Experimental Gerontology, 27*(2), 241—250. https://doi.org/10.1016/0531-5565(92)90048-5

Service, P. M. (1987). Physiological mechanisms of increased stress resistance in *Drosophila melanogaster* selected for postponed senescence. *Physiological Zoology, 60*(3), 321—326.

Service, P. M., Hutchinson, E. W., MacKinley, M. D., & Rose, M. R. (1985). Resistance to environmental stress in *Drosophila melanogaster* selected for postponed senescence. *Physiological Zoology, 58*(4), 380—389.

CHAPTER FORTY FIVE

1996: Late-life mortality plateaus explained using evolutionary theory

The standard paradigm

The conventional way of modeling adult demography across disciplines is to assume an exponential increase in adult mortality with age. Exponential terms are indeed incorporated in the evolutionary theory developed for age-structured populations, from Norton (1928) to Charlesworth (1980). Even the forces of natural selection derived by Hamilton (1966) incorporate such exponential terms.

But more extensive human demographic data often fail to conform to the Gompertz model at very advanced ages. Historical examples of such demographic data include the mortality data modeled in Greenwood and Irwin (1939), but such data were viewed as the exception not the rule. Later demographers, however, considered the possibility that deviations from Gompertz patterns were important and consequential, rather than anomalies to be ignored. Like Greenwood and Irwin (1939), Beard (1959) and Vaupel et al. (1979) proposed that the deceleration in mortality rates at later adult ages might be explained in terms of lifelong heterogeneity for robustness.

As described in Chapter 38, Carey et al. (1992) and Curtsinger et al. (1992) provided experimental evidence for the existence of mortality plateaus occurring in advanced ages of laboratory populations, which further challenged Gompertzian models. These findings provoked evolutionary biologists to consider how mortality models *should* behave, in light of explicit evolutionary models for life history at very late ages in populations with age-structure. To address this question, it was necessary to determine if such mortality plateaus were indeed in congruence with the conventional evolutionary explanation of aging.

Conceptual Breakthroughs in The Evolutionary Biology of Aging
ISBN: 978-0-12-821545-6
https://doi.org/10.1016/B978-0-12-821545-6.00033-9

The conceptual breakthrough

The aim of Mueller and Rose (1996) was to show that late-life mortality plateaus were not only in agreement with the evolutionary explanation of aging, but that they provided deep insights into the genetic mechanisms of life-history evolution. Using numerical simulations of life-history evolution in clonal populations, Mueller and Rose (1996) showed that late-life mortality rate plateaus evolved using mutations both with and without antagonistic pleiotropy. Subsequent mathematical analysis based chiefly on the mutation accumulation hypothesis likewise showed that evolution produced late-life mortality plateaus without any need for lifelong heterogeneity effects (Wachter, 1999; Charlesworth, 2001).

The simulations of Mueller and Rose (1996) assumed initial mortality patterns conforming to the standard Gompertz mortality model. They then introduced a variety of mutations with small effects to reshape the mortality pattern. Although there are multiple ways in which new mutations might be introduced into a simulated population, with variations in their degree of similarities to the host population, Mueller and Rose (1996) decided to use the continuum-of-mutation model to minimize departures that would be too easily weeded out by natural selection. In this way, new mutants would have a higher chance of improving fitness by minimizing both risk and gains to population fitness. The simulation assumed a fixed maximum lifespan of 109 and a 9-day development time, giving 100 discrete adult age classes and nine juvenile classes to model the dipteran fly life cycle. Given these features, all their simulations of life-history evolution led to the eventual production of late-life mortality plateaus, albeit plateaus that were not necessarily perfectly flat. Their simulated plateaus often still had a slight upward trend toward higher mortality rates, but that upward trend was quite slight compared to the continued exponential rise in mortality rates inherent in Gompertzian demographic models (Mueller & Rose, 1996).

Impact: 9

These results were criticized by Pletcher and Curtsinger (1998) and Wachter (1999), who argued that mutation accumulation and antagonistic pleiotropy did not robustly explain mortality plateaus. Specifically, Pletcher, Curtsinger, and Wachter called into question the use of mutants that only affect one age class in some of the numerical calculations of Mueller and Rose (1996). Nonetheless, it had at least been established that late-life mortality rate plateaus were not anomalous for the evolutionary theory of aging developed by Hamilton (1966) and Charlesworth (1980). In particular, assuming without evidence extreme levels of lifelong robustness was not necessary to explain the cessation of aging, at least in theory. Mueller et al. (2011) would take on criticisms of this kind, as well as deeply examining the alternative lifelong heterogeneity model.

Without specifically addressing such criticisms at this point, it is important to point out that the results of Mueller and Rose (1996) established that late-life plateaus in mortality were not falsifications of the evolutionary theory of aging. They were, instead, potential corroborations of an expanded theory for the evolution of life-history before, during, and notably *after reproductive ages*. The continued development of that expanded theory would be a major concern for the next 15 years. At least as importantly, strong-inference experiments testing that expanded theory have been a major feature of research on the evolutionary biology of aging ever since. Thus this conceptual breakthrough was the most important turning point in the development of the evolutionary biology of aging since the 1980 book of Charlesworth.

References and further reading

Beard, R. E. (1959). Note on some mathematical mortality models. The lifespan of animals. In G. E. W. Wolstenholme, & M. O'Conner (Eds.), *Ciba foundation colloquium on ageing* (pp. 302–311). Boston: Little Brown and Company. https://doi.org/10.1002/9780470715253.app1

Carey, J. R., Liedo, P., Orozco, D., & Vaupel, J. W. (1992). Slowing of mortality rates at older ages in large medfly cohorts. *Science, 258*(5081), 457–461. (JSTOR).

Charlesworth, B. (1980). *Evolution in age-structured populations.* Cambridge: Cambridge University Press.

Charlesworth, B. (2001). Patterns of age-specific means and genetic variances of mortality rates predicted by the mutation-accumulation theory of ageing. *Journal of Theoretical Biology, 210*(1), 47–65. https://doi.org/10.1006/jtbi.2001.2296

Curtsinger, J. W., Fukui, H. H., Townsend, D. R., & Vaupel, J. W. (1992). Demography of genotypes: Failure of the limited life-span paradigm in *Drosophila melanogaster. Science, 258*(5081), 461–463. https://doi.org/10.1126/science.1411541

Gompertz, B. (1825). On the nature of the function expressive of the law of human mortality, and on a new mode of determining the value of life contingencies. In F. R. S. Esq (Ed.), *A letter to Francis Baily* (Vol. 115, pp. 513–583). Philosophical Transactions of the Royal Society of London. https://doi.org/10.1098/rstl.1825.0026

Greenwood, M., & Irwin, J. O. (1939). The biostatistics of senility. *Human Biology, 11*(1), 1–23.

Hamilton, W. D. (1966). The moulding of senescence by natural selection. *Journal of Theoretical Biology, 12*(1), 12–45. https://doi.org/10.1016/0022-5193(66)90184-6

Kirkwood, T. B. L. (2015). Deciphering death: A commentary on Gompertz (1825) on the nature of the function expressive of the law of human mortality, and on a new mode of determining the value of life contingencies. *Philosophical Transactions of the Royal Society B: Biological Sciences, 370*(1666), 20140379. https://doi.org/10.1098/rstb.2014.0379

Mueller, L. D., Rauser, C. L., & Rose, M. R. (2011). *Does aging stop?* Oxford: Oxford University Press.

Mueller, L. D., & Rose, M. R. (1996). Evolutionary theory predicts late-life mortality plateaus. *Proceedings of the National Academy of Sciences of the United States of America, 93*(26), 15249–15253.

Norton, H. T. J. (1928). Natural selection and mendelian variation. *Proceedings of the London Mathematical Society, s2–28*(1), 1–45. https://doi.org/10.1112/plms/s2-28.1.1

Pletcher, S. D., & Curtsinger, J. W. (1998). Mortality plateaus and the evolution of senescence: Why are old-age mortality rates so low? *Evolution, 52*(2), 454–464. https://doi.org/10.2307/2411081. JSTOR.

Vaupel, J. W., Manton, K. G., & Stallard, E. (1979). The impact of heterogeneity in individual frailty on the dynamics of mortality. *Demography, 16*(3), 439–454.

Wachter, K. W. (1999). Evolutionary demographic models for mortality plateaus. *Proceedings of the National Academy of Sciences of the United States of America, 96*(18), 10544–10547.

CHAPTER FORTY SIX

1998–2003: Falsification of lifelong heterogeneity models for the cessation of aging

The standard paradigm

The observation of decelerating mortality rates at advanced ages in age-synchronized laboratory cohorts of both inbred and outbred organisms motivated the development of hypotheses to explain this phenomenon. Outside of the evolutionary biology of aging, the chief hypothesis that was used to explain these mortality plateaus was lifelong heterogeneity, as previously mentioned (See Greenwood and Irwin (1939) or Vaupel et al. (1979), for examples). With lifelong and extreme heterogeneity for robustness, less robust individuals with rapid aging should be largely eliminated from individual same-aged cohorts by sufficiently late ages, leaving only the highly robust individuals with much lower mortality rates. With such individuals dominating same-aged cohorts, mortality rates will thus approximately plateau at late ages. In a sense, on this hypothesis, aging continues in a Gompertzian fashion. Aging only stops with respect to aggregate mortality rates, at a cohort-level, according to the lifelong heterogeneity model.

The evolutionary theory of late-life mortality deceleration (e.g., Charlesworth, 2001; Mueller & Rose, 1996) explains mortality rate plateauing as a feature of many, if not most, individuals in a cohort. That is to say, the aging of individual members of cohorts actually slows and comes close to stopping. In effect, what might be called "individual aging" actually stops, on this hypothesis.

A third hypothesis that was brought forward to explain late-life mortality-rate plateaus was experimental artifacts (e.g., Nusbaum et al., 1993). The Curtsinger lab performed a number of studies which showed that the proposed artifacts were unlikely to generate the mortality-rate plateaus observed by Carey et al. (1992) and Curtsinger et al. (1992). For example, Fukui et al. (1993) extended the work of Curtsinger et al. (1992), showing that mortality rate plateaus were consistently observed in both sexes across four different inbred stocks of *D. melanogastger*. Khazaeli

Conceptual Breakthroughs in The Evolutionary Biology of Aging
ISBN: 978-0-12-821545-6
https://doi.org/10.1016/B978-0-12-821545-6.00058-3

et al. (1995) showed that mortality plateaus were still observed when cohort densities were varied in *D. melanogaster*. These publications and others like them firmly established that further research had to decide between the lifelong heterogeneity and evolutionary hypotheses that might, in principle, explain late-life mortality plateaus.

The conceptual breakthrough

The conventional demographic explanation of the cessation of aging in terms of lifelong heterogeneity requires very extreme differences in robustness among members of a cohort. Fluctuations in environmental or genetic heterogeneity should therefore lead to major differences in patterns of aging cessation. Yet both inbred and outbred lines show similar decelerations in mortality rate increases, with comparable cessations of aging. This suggested that attempts to reduce individual heterogeneity should diminish mortality-rate deceleration, if the lifelong heterogeneity hypothesis was the best explanation of these plateaus. Furthermore, since the lifelong heterogeneity model rested specifically on unproven assumptions about robustness, major changes in robustness in experimental populations should alter the timing or height of the mortality plateaus.

The first strong test of the lifelong heterogeneity hypothesis was performed by Khazaeli et al. (1998), who treated numerous *D. melanogaster* cohorts either in a heterogeneity-increasing manner or in a heterogeneity-decreasing manner. In an attempt to reduce environmental variation, meticulous effort was devoted toward making identical environments for every larva. This was accomplished by imposing a tight window for collecting eggs, controlling the exact number of larvae per vial, and collecting flies from pupae of similar positioning. In this way, the flies involved were not only highly inbred, but also subjected to as identical an environment as possible. Yet it was found that this manipulation had no significant impact on the mortality rate plateaus that they observed (Khazaeli et al., 1998).

Using very different methods, Drapeau et al. (2000) compared the mortality rate plateaus of 15 *D. melanogaster* populations that had evolved highly differentiated levels of resistance to starvation stress, thus featuring strikingly different levels of lifelong robustness. But despite these substantial differences in adult robustness, these populations did not differ with respect to their mortality-rate plateaus (It should be noted that these populations had similar declines in the forces of natural selection. This will prove significant in the tests of the evolutionary theory of late life, discussed in Chapter 49).

Mueller et al. (2003) developed an explicit mathematical model of lifelong heterogeneity and its effects on cohort-level mortality. Then they used data collected from fruit fly populations that exhibited mortality-rate plateaus to determine the model parameters required for lifelong heterogeneity to produce the mortality-rate plateaus of those populations. Note that this was an attempt to force a lifelong heterogeneity model to work. But the parameter values required to make lifelong heterogeneity models work were far outside the range of observed variation in lifespan found in any of these *D. melanogaster* populations.

Impact: 8

All three of these early tests of the lifelong heterogeneity model falsified the hypothesis. Further experimental tests summarized in Mueller et al. (2011), and discussed in Chapter 45, similarly falsified the hypothesis. Though the lifelong heterogeneity model has the same intuitive appeal as the geocentric model of astrophysics, it has repeatedly failed to survive strong-inference tests. What remained to be determined was whether the evolutionary hypothesis for the deceleration of aging late in adult life was equally flawed.

References and further reading

Carey, J. R., Liedo, P., Orozco, D., & Vaupel, J. W. (1992). Slowing of mortality rates at older ages in large medfly cohorts. *Science, 258*(5081), 457−461 (JSTOR).

Charlesworth, B. (2001). Patterns of age-specific means and genetic variances of mortality rates predicted by the mutation-accumulation theory of ageing. *Journal of Theoretical Biology, 210*(1), 47−65. https://doi.org/10.1006/jtbi.2001.2296

Curtsinger, J. W., Fukui, H. H., Townsend, D. R., & Vaupel, J. W. (1992). Demography of genotypes: Failure of the limited life-span paradigm in *Drosophila melanogaster*. *Science, 258*(5081), 461−463. https://doi.org/10.1126/science.1411541

Drapeau, M. D., Gass, E. K., Simison, M. D., Mueller, L. D., & Rose, M. R. (2000). Testing the heterogeneity theory of late-life mortality plateaus by using cohorts of *Drosophila melanogaster*. *Experimental Gerontology, 35*(1), 71−84. https://doi.org/10.1016/S0531-5565(99)00082-0

Fukui, H. H., Xiu, L., & Curtsinger, J. W. (1993). Slowing of age-specific mortality rates in *Drosophila melanogaster*. *Experimental Gerontology, 28*(6), 585−599. https://doi.org/10.1016/0531-5565(93)90048-I

Gompertz, B. (1825). On the nature of the function expressive of the law of human mortality, and on a new mode of determining the value of life contingencies. In F. R. S. Esq (Ed.), *A letter to Francis Baily* (Vol. 115, pp. 513−583). Philosophical Transactions of the Royal Society of London. https://doi.org/10.1098/rstl.1825.0026

Greenwood, M., & Irwin, J. O. (1939). The biostatistics of senility. *Human Biology, 11*(1), 1–23.

Khazaeli, A. A., Pletcher, S. D., & Curtsinger, J. W. (1998). The fractionation experiment: Reducing heterogeneity to investigate age-specific mortality in Drosophila. *Mechanisms of Ageing and Development, 105*(3), 301–317. https://doi.org/10.1016/S0047-6374(98)00102-X

Khazaeli, A. A., Xiu, L., & Curtsinger, J. W. (1995). Effect of adult cohort density on age-specific mortality in *Drosophila melanogaster*. *The Journals of Gerontology Series A: Biological Sciences and Medical Sciences, 50A*(5), B262–B269. https://doi.org/10.1093/gerona/50A.5.B262

Mueller, L. D., et al. (2003). Statistical tests of demographic heterogeneity theories. *Experimental Gerontology, 38*(4), 373–386. https://doi.org/10.1016/s0531-5565(02)00238-3

Mueller, L. D., Nusbaum, T. J., & Rose, M. R. (1995). The Gompertz equation as a predictive tool in demography. *Experimental Gerontology, 30*(6), 553–569. https://doi.org/10.1016/0531-5565(95)00029-1

Mueller, L. D., Rauser, C. L., & Rose, M. R. (2011). *Does aging stop?* Oxford University Press. https://doi.org/10.1093/acprof:oso/9780199754229.001.0001

Mueller, L. D., & Rose, M. R. (1996). Evolutionary theory predicts late-life mortality plateaus. *Proceedings of the National Academy of Sciences of the United States of America, 93*(26), 15249–15253.

Nusbaum, T. J., Graves, J. L., Mueller, L. D., & Rose, M. R. (1993). Fruit fly aging and mortality. *Science, 260*(5114), 1567–1569. https://doi.org/10.1126/science.8503001

Vaupel, James W., Manton, Kenneth G., & Stallard, Eric (1979). The impact of heterogeneity in individual frailty on the dynamics of mortality. *Demography, 16*(3), 439–454.

CHAPTER FORTY SEVEN

1998—2000: Discovery of Drosophila mutants that sometimes increase longevity

The standard paradigm

A central feature of the reductionist cell-molecular paradigm for aging is that it should be possible to find large-effect mutations that alleviate a physiological deficit in cell metabolism, and thereby substantially increase longevity. This search for large-effect beneficial mutations falls in line with the principles of "classical" evolutionary genetic theory, which presumed low levels of segregating Mendelian variation (Lewontin, 1974). Once introduced into a population, this classical version of evolutionary theory supposed, these rare beneficial mutations would sweep to fixation.

Starting in the 1960s, the classical theory was undermined by overwhelming evidence for widespread genetic variation in Mendelian populations (Lewontin, 1974). Subsequent genomic research revealed so much segregating genetic variation in outbred animal populations that the classical evolutionary genetic model is only tenable for strictly asexual populations, or sexual populations that very rarely undergo recombination. This motivated the replacement of classical theory with present day neoclassical evolutionary-genetic theory. In neoclassical theory, abundant segregating genetic variation genome-wide is explained in terms of two types of evolutionary genetic features: neutral genetic variation due to drift; and weakly deleterious mutations that are maintained at low frequencies by selection-mutation balance. But the classical assumption that the genetic foundations of adaptation are rare beneficial mutations that quickly sweep to fixation is maintained (Burke, 2012).

The discovery and characterization of nematode mutants that greatly increase longevity (e.g., Kenyon, 1993; Van Voorhies, 1992) was consistent with neoclassical evolutionary genetic theory, because *C. elegans* rarely undergoes outcrossing. Most reproduction in this species of nematode occurs when hermaphrodites self-fertilize. Longevity mutants found in organisms that experience recombination only rarely are more likely to improve fitness

Conceptual Breakthroughs in The Evolutionary Biology of Aging
ISBN: 978-0-12-821545-6
https://doi.org/10.1016/B978-0-12-821545-6.00009-1

in a more global sense, improving both survival and reproduction as they adapt to a novel laboratory environment. The question that remained was whether similar mutants existed in Drosophila, mutants that would strikingly increase lifespan.

The conceptual breakthrough

The first Drosophila mutant that seemed to dramatically increase lifespan was named *methuselah* by its discoverers (Lin et al., 1998). The P-element insertion mutant *mth* displayed an improved average longevity relative to its parent line of $\sim 35\%$. It was also found to improve resistance to such stressors as starvation, heat shock, and an artificial diet laced with free-radicals. The *mth* gene encodes a G-protein coupled receptor protein that is part of a signal transduction pathway which modulates stress resistance, development, and longevity.

The next supposed Drosophila longevity gene was called *indy*, an acronym for "I'm not dead yet," a line from a Monty Python film (Rogina et al., 2000). The result of five independent P-element insertions, *indy* was found to yield a total lifespan twice that of its control and revertant populations. Rogina et al. (2000) believed that the *indy* protein functions as a "sodium dicarboxylate cotransporter" in mammals, regulating the Krebs cycle and thus intermediary metabolism. Rogina et al. (2000) were able to localize the gene's expression to the midgut, fat bodies, and oenocytes of their flies.

To this point in the study of Drosophila longevity mutants, the results seemed to match those found for nematode longevity mutants. But problems with these mutants surfaced in research conducted by Khazaeli et al. (2005). In their study, Khazaeli et al. (2005) sought to find the relationship between body size, mass, and life span between varying lines of outbred, recombinant inbred, inbred and mutant lines of *Drosphila melanogaster*. The "longevity" mutants, *mth* and *indy*, didn't consistently live as long as their short-lived ancestral w^{1118} stock, suggesting that their longevity extension was not readily reproducible. This failure of replication could have been due to small variations in assay conditions combined with genotype-by-environment interaction. Alternatively, the inbreeding depression that is so common in Drosophila culture may have undermined the benefits of these longevity mutants, because of inbreeding depression [As a consistently self-fertilizing species, *C. elegans* are naturally highly inbred, and thus deleterious recessive mutations are rapidly eliminated by selection against such homozygotes, thus obviating inbreeding depression]. And it is possible that both of these artifactual problems could have undermined the longevity benefits of these Drosophila mutations.

As a positive control on the quality of the work by Khazaeli et al. (2005), their assays of the stocks developed by Rose (1984b) using experimental evolution robustly displayed increased longevity. Indeed, those stocks featured by far the greatest longevities among all the stocks studied by Khazaeli et al. (2005).

Impact: 3

While longevity mutants obtained in self-fertilizing nematodes show durable extensions of lifespan, ostensibly similar Drosophila mutants do not. It is difficult to be certain which particular genetic artifact undermines the value of these Drosophila longevity mutants. But it is evident that they are not reliable experimental material. We have already discussed the challenges of inbreeding depression and genotype-by-environment in previous chapters; in the present case, we are invoking them in order to explain why the initial work with these Drosophila mutants could be valid, but then other labs were unable to reproduce the initial findings.

By contrast, populations of outbred Drosophila that have increased lifespan as a result of experimental evolution (e.g., Luckinbill et al., 1984) are more reliably differentiated, so long as they are not deliberately or inadvertently inbred (vid. Rose, 1984a). Additional progress on the physiology, genomics, and transcriptomics of aging Drosophila would depend chiefly on outbred stocks of such longer-lived populations.

References and further reading

Burke, M. K. (2012). How does adaptation sweep through the genome? Insights from long-term selection experiments. *Proceedings of the Royal Society B: Biological Sciences, 279*(1749), 5029–5038. https://doi.org/10.1098/rspb.2012.0799

Kenyon, C., Chang, J., Gensch, E., Rudner, A., & Tabtiang, R. (1993). A *C. elegans* mutant that lives twice as long as wild type. *Nature, 366*(6454), 461–464. https://doi.org/10.1038/366461a0

Khazaeli, A., Vanvoorhies, W., & Curtsinger, J. (2005). The relationship between life span and adult body size is highly strain-specific in. *Experimental Gerontology, 40*(5), 377–385. https://doi.org/10.1016/j.exger.2005.02.004

Lewontin, R. C. (1974). *The genetic basis of evolutionary change.* Columbia University Press.

Lin, Y.-J., Seroude, L., & Benzer, S. (1998). Extended life-span and stress resistance in the Drosophila mutant methuselah. *Science, 282*(5390), 943–946 (JSTOR).

Luckinbill, L. S., Arking, R., Clare, M. J., Cirocco, W. C., & Buck, S. A. (1984). Selection for delayed senescence in *Drosophila melanogaster. Evolution, 38*(5), 996–1003. https://doi.org/10.2307/2408433. JSTOR.

Rogina, B., Reenan, R. A., Nilsen, S. P., & Helfand, S. L. (2000). Extended life-span conferred by cotransporter gene mutations in Drosophila. *Science, 290*(5499), 2137–2140 (JSTOR).

Rose, M. R. (1984a). Genetic covariation in Drosophila life history: Untangling the data. *The American Naturalist, 123*(4), 565–569. https://doi.org/10.1086/284222

Rose, M. R. (1984b). Laboratory evolution of postponed senescence in *Drosophila melanogaster. Evolution, 38*(5), 1004–1010. https://doi.org/10.2307/2408434. JSTOR.

Van Voorhies, W. A. (1992). Production of sperm reduces nematode lifespan. *Nature, 360*(6403), 456–458. https://doi.org/10.1038/360456a0

CHAPTER FORTY EIGHT

1999—2004: Nematode longevity mutants show antagonistic pleiotropy

The standard paradigm

In keeping with the cell-molecular paradigm for aging, some nematode geneticists (e.g., Cypser & Johnson, 1999; Kenyon et al., 1993) argued that the mutants with increased longevity are even more healthy than nematodes that don't have them. They have characterized such mutants as "longevity assurance" genes that give individuals bearing them a longer lifespan because of improved robustness, vigor, etc. Johnson (1990) and Kenyon et al. (1993) seemed to demonstrate that the *age-1* mutant and other *daf* regulatory mutants provided global benefits that were broadly beneficial. For example, they argued that a molecular mechanism responsible for the antiaging properties of these mutant genes was enhanced repair of metabolic damage (Martin et al., 1996; Lithgow & Kirkwood, 1996).

This claim of generally beneficial effects was always amenable to empirical test. What was required was attention to functional characters that might be reduced in the longer-lived nematode mutants, potentially under a subset of relevant environmental conditions.

The conceptual breakthrough

Multiple studies demonstrated that the life-extending qualities of "longevity assurance" mutant *C. elegans* were context dependent (Again, another instance where genotype-by-environment interaction might have played a role). Changing such environmental conditions as food availability and developmental temperatures sometimes led to results that did not fit with prior characterizations of these mutants as broadly beneficial. Work on the metabolic rate of dauer-gene longevity mutants by Van Voorhies and Ward (1999) showed that these longevity mutants have reduced

Conceptual Breakthroughs in The Evolutionary Biology of Aging
ISBN: 978-0-12-821545-6
https://doi.org/10.1016/B978-0-12-821545-6.00030-3

metabolic rates, suggesting that such mutants have dauer-like physiological properties as adults. By comparing the standard N2 strain with "long-lived" mutants like the original *age-1* and *daf-2-(e1370)*, Van Voorhies and Ward (1999) measured mean longevity and metabolic rates with varying temperature. What they found was increased longevity occurred only in association with reduced metabolic rate, suggesting a trade-off between the two characteristics, rather than a broadly adaptive aging benefit from having the mutant. In effect, the work of Van Voorhies and Ward (1999) appears to reveal some type of antagonistic pleiotropy in the genetics of nematode aging.

Walker et al. (2000) and Jenkins et al. (2004) found that nematode longevity mutants were selected against when they were forced to compete with their own ancestral strains. Walker et al. (2000) compared wild-type *C. elegans* with *age-1(hx546)* mutant worms, finding a similar temperature-dependent expression of dauer qualities in the mutants.

Van Voorhies et al. (2006) thus proposed that all mutants which show benefits for aging under some conditions, for some characters, will eventually be found to have antagonistic pleiotropic effects on Darwinian fitness, especially when assayed in their ancestral environments (It is of note that Leroi et al. (1994) found a similar pattern of environment-dependence for the trade-offs exhibited by longer-lived Drosophila produced by experimental evolution).

Impact: 8

By the early 2000s, it was growing apparent that evolutionary mechanisms and considerations were too important to be neglected by the geneticists who studied animals like nematodes and fruit flies. Mendelian and molecular geneticists who studied aging were not usually interested in evolutionary constraints like genotype-by-environment interaction, inbreeding depression, and antagonistic pleiotropy. But such evolutionary constraints demonstrably impinged on their research, nonetheless. By this point in the development of the field of gerontology as a whole, the persistent neglect of these well-established challenges facing the study of aging was undermining the quality of the research conducted by those gerontologists who were seemingly determined to deny the evolutionary foundations of aging.

References and further reading

Cypser, J. R., & Johnson, T. E. (1999). The spe-10 mutant has longer life and increased stress resistance. *Neurobiology of Aging, 20*(5), 503—512. https://doi.org/10.1016/S0197-4580(99)00085-8

Jenkins, N. L., McColl, G., & Lithgow, G. J. (2004). Fitness cost of extended lifespan in *Caenorhabditis elegans*. *Proceedings of the Royal Society B: Biological Sciences, 271*(1556), 2523—2526. https://doi.org/10.1098/rspb.2004.2897

Johnson, T. E. (1990). Increased life-span of *age* -1 mutants in *Caenorhabditis elegans* and lower gompertz rate of aging. *Science, 249*(4971), 908—912. https://doi.org/10.1126/science.2392681

Kenyon, C., Chang, J., Gensch, E., Rudner, A., & Tabtiang, R. (1993). A *C. elegans* mutant that lives twice as long as wild type. *Nature, 366*(6454), 461—464. https://doi.org/10.1038/366461a0

Leroi, A. M., Chippindale, A. K., & Rose, M. R. (1994). Long-term laboratory evolution of a genetic life-history trade-off in *Drosophila melanogaster*. 1. The role of genotype-by-environment interaction. *Evolution, 48*(4), 1244—1257. https://doi.org/10.2307/2410382. JSTOR.

Lithgow, G. J., & Kirkwood, T. B. L. (1996). Mechanisms and evolution of aging. *Science, 273*(5271), 80. https://doi.org/10.1126/science.273.5271.80

Martin, G. M., Austad, S. N., & Johnson, T. E. (1996). Genetic analysis of ageing: Role of oxidative damage and environmental stresses. *Nature Genetics, 13*(1), 25—34. https://doi.org/10.1038/ng0596-25

Van Voorhies, W. A., Curtsinger, J. W., & Rose, M. R. (2006). Do longevity mutants always show trade-offs? *Experimental Gerontology, 41*(10), 1055—1058. https://doi.org/10.1016/j.exger.2006.05.006

Van Voorhies, W. A., & Ward, S. (1999). Genetic and environmental conditions that increase longevity in *Caenorhabditis elegans* decrease metabolic rate. *Proceedings of the National Academy of Sciences, 96*(20), 11399—11403. https://doi.org/10.1073/pnas.96.20.11399

Walker, D. W., McColl, G., Jenkins, N. L., Harris, J., & Lithgow, G. J. (2000). Evolution of lifespan in *C. elegans*. *Nature, 405*(6784), 296. https://doi.org/10.1038/35012693

CHAPTER FORTY NINE

2002—06: Evolution of life-history fits evolutionary analysis of late life

The standard paradigm

The two major explanations for the stabilization of late-life mortality rates at advanced ages are lifelong heterogeneity in robustness or the force of natural selection plateauing after the end of reproduction. If conventional lifelong heterogeneity theories (Vaupel et al., 1979; Vaupel, 1988, 1993,1998) for the cessation of aging are correct, then experimental evolution of late life should not match the predictions of the 1996 Mueller-Rose evolutionary theory. That evolutionary theory, discussed in greater detail in Chapter 46, hinges on the onset of mortality-rate plateaus arising from the late-life plateaus in the forces of natural selection. That in turn implies that manipulating the timing of the cessation of reproduction during a population's evolution should ultimately lead to shifts in the onset of its mortality-rate plateaus, with later cessation of reproduction leading to later mortality plateaus, for example.

Furthermore, this evolutionary logic applies equally well to the evolution of age-specific fecundity. Since the force of natural selection acting on age-specific fecundity also declines to a plateau in late life, so long as populations are not declining rapidly toward extinction, then age-specific fecundity should also have a tendency to plateau at very late ages. But no such effect is inherent to the lifelong heterogeneity theory as originally proposed.

The conceptual breakthrough

The aim of Rose et al. (2002) was to test the evolutionary explanation for late life mortality-rate plateaus using populations of outbred lines of *Drosophila melanogaster* that had different ages at which population reproduction was terminated. They used computer simulations of the evolution of populations with different last ages of reproduction to demonstrate that their

Conceptual Breakthroughs in The Evolutionary Biology of Aging
ISBN: 978-0-12-821545-6
https://doi.org/10.1016/B978-0-12-821545-6.00023-6

Hamiltonian logic would work numerically: the age at which mortality plateaus demonstrably depends on the last age at which a population reproduces. That work implies that, if the age at which mortality rates plateau is the same regardless of the last age of reproduction over many generations, then the evolutionary explanation of such plateaus would be falsified.

Rose et al. (2002) performed two independent tests of this prediction, using the Drosophila populations of Rose (1984) and the Drosophila populations of Chippindale et al. (1997). In both cases, they found that the mortality plateaus shifted in accordance with the evolutionary theory of late life, which they had explicitly simulated. Further tests were performed to determine the genetic mechanism(s) underlying the evolution of late-life mortality, whether mutation accumulation, antagonistic pleiotropy, or both. By testing for hybrid vigor between crosses of genetically divergent lines, Rose et al. (2002) explored the potential of mutation accumulation involving recessive deleterious mutations with effects at later ages. But the mortality plateaus of hybrid cohorts exhibited no significant difference from their uncrossed parental cohorts, providing no evidence for that kind of mutation accumulation (Nonetheless, other types of mutations, such as strictly additive deleterious mutations, could have accumulated in these populations).

By contrast, there was strong evidence for antagonistic pleiotropy from reverse evolution experiments performed by switching derivatives of the longer-lived O populations onto an evolutionary regime that featured early reproduction, the "NRO" populations. After just dozens of generations of reversed selection, these NRO populations had an earlier onset of mortality-rate plateaus. This rapid shift in mortality rates also provided further evidence for a negative trade-off between longevity and early reproduction, lending further credence to the antagonistic pleiotropy mechanism for the evolution of aging generally.

Rauser et al. (2005) developed theory for the evolution of aging cessation among reproductive characters, predicting that such cessations should arise for both age-specific fecundity and age-specific virility (See Fig. 49.1). Such cessations were found in Drosophila by Rauser et al. (2003) for fertility and by Shahrestani et al. (2012) for virility.

The aim of Rauser et al. (2006) was to test whether the evolutionary theories of late-life (Charlesworth, 2001; Mueller & Rose, 1996; Rose & Mueller, 2000) apply to the laboratory evolution of fecundity as predicted by Rauser et al. (2005). The fecundity assays of Rauser et al. (2006) showed that the ages of onset of Drosophila fecundity plateaus were different

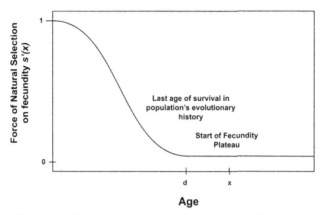

Figure 49.1 The force of natural selection acting on age-specific fecundity. "d" represents the last age of survival in the population's history, while "x" marks the beginning of the fecundity plateau after the force of natural selection is 0. *Inspired by Rauser, C. L., Abdel-Aal, Y., Shieh, J. A., Suen, C. W., Mueller, L. D., and Rose, M. R. (2005). Lifelong heterogeneity in fecundity is insufficient to explain late-life fecundity plateaus in* Drosophila melanogaster. *Experimental Gerontology, 40(8), 660–670. https://doi.org/10.1016/j.exger.2005.06.006*

between five early cultured ACO populations and five late-culured CO's, showing that fecundity plateaus too are dependent on when the force of natural selection on fecundity falls to zero. In addition to supporting the evolutionary explanation for late life, reverse selection again supported the action of antagonistic pleiotropy. Reverse-selection NRCO's were subjected to early culture reproduction for 24 generations. When the NRCO's were compared to their ancestral CO lines, as shown in Fig. 49.2, they had an earlier fecundity plateau, suggesting pleiotropic interconnection between early fitness-components and later fecundity.

Impact: 10

All told, these experiments showed that aging stops at later ages in cohorts taken from populations that have later *last ages* of reproduction in laboratory cultures. This was an indispensable corroboration of the evolutionary theory of late life, indicating that evolutionary theories which depend on the forces of natural selection can account for the cessation of aging observed by Carey and Curtsinger, as described in Chapter 38. Notably, if these experiments had *not* produced such results, the evolutionary theory of late life would have been falsified.

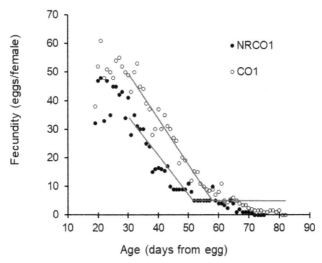

Figure 49.2 Mean fecundity as a function of age for each of NRCO1 (early reproducing) and CO1 (late reproducing) populations. The NRCO populations generally exhibited an earlier onset of the late-life fecundity plateaus relative to their ancestral CO lines. This provided strong evidence for antagonistic pleiotropy as a genetic mechanism working to shape late age fecundity patterns. *Data from Rauser, C. L., Abdel-Aal, Y., Shieh, J. A., Suen, C. W., Mueller, L. D., and Rose, M. R. (2005). Lifelong heterogeneity in fecundity is insufficient to explain late-life fecundity plateaus in* Drosophila melanogaster. *Experimental Gerontology, 40(8), 660–670. https://doi.org/10.1016/j.exger.2005.06.006*

This was an unusually successful turn of events for any theory in biology: the ostensibly anomalous phenomenon of late-life mortality plateaus turned out to be explicable in terms of the evolutionary theory originally developed only to explain aging, not its cessation. Furthermore, previously unobserved plateaus in reproductive characters were predicted from the evolutionary theory of aging, and then discovered. Finally, both types of life-history character proved to be evolutionarily modifiable in a reproducible manner, as predicted by extensions of the original theory.

References and further reading

Charlesworth, B. (2001). Patterns of age-specific means and genetic variances of mortality rates predicted by the mutation-accumulation theory of ageing. *Journal of Theoretical Biology, 210*(1), 47–65. https://doi.org/10.1006/jtbi.2001.2296

Chippindale, A. K., Alipaz, J. A., Chen, H.-W., & Rose, M. R. (1997). Experimental evolution of accelerated development in Drosophila. 1. Developmental speed and larval survival. *Evolution, 51*(5), 1536–1551. https://doi.org/10.1111/j.1558-5646.1997.tb01477.x

Mueller, L. D., & Rose, M. R. (1996). Evolutionary theory predicts late-life mortality plateaus. *Proceedings of the National Academy of Sciences of the United States of America, 93*(26), 15249—15253.

Rauser, C. L., Abdel-Aal, Y., Shieh, J. A., Suen, C. W., Mueller, L. D., & Rose, M. R. (2005). Lifelong heterogeneity in fecundity is insufficient to explain late-life fecundity plateaus in *Drosophila melanogaster. Experimental Gerontology, 40*(8), 660—670. https://doi.org/10.1016/j.exger.2005.06.006

Rauser, C. L., Mueller, L. D., & Rose, M. R. (2003). Aging, fertility, and immortality. *Experimental Gerontology, 38*(1), 27—33. https://doi.org/10.1016/S0531-5565(02)00148-1

Rauser, C. L., Tierney, J. J., Gunion, S. M., Covarrubias, G. M., Mueller, L. D., & Rose, M. R. (2006). Evolution of late-life fecundity in *Drosophila melanogaster. Journal of Evolutionary Biology, 19*(1), 289—301. https://doi.org/10.1111/j.1420-9101.2005.00966.x

Rose, M. R., Dorey, M. L., Coyle, A. M., & Service, P. M. (1984). The morphology of postponed senescence in *Drosophila melanogaster. Canadian Journal of Zoology, 62*(8), 1576—1580. https://doi.org/10.1139/z84-230

Rose, M. R., Drapeau, M. D., Yazdi, P. G., Shah, K. H., Moise, D. B., Thakar, R. R., Rauser, C. L., & Mueller, L. D. (2002). Evolution of late-life mortality in *Drosophila melanogaster. Evolution, 56*(10), 1982—1991. https://doi.org/10.1111/j.0014-3820.2002.tb00124.x

Rose, M. R., & Mueller, L. D. (2000). Ageing and immortality. *Philosophical Transactions of the Royal Society of London. Series B: Biological Sciences, 355*(1403), 1657—1662. https://doi.org/10.1098/rstb.2000.0728

Shahrestani, P., Tran, X., & Mueller, L. D. (2012). Patterns of male fitness conform to predictions of evolutionary models of late life: Late-life male fitness. *Journal of Evolutionary Biology, 25*(6), 1060—1065. https://doi.org/10.1111/j.1420-9101.2012.02492.x

Vaupel, J. W. (1988). Inherited frailty and longevity. *Demography, 25*(2), 277—287. https://doi.org/10.2307/2061294

Vaupel, J. W., & Carey, J. R. (1993). Compositional interpretations of medfly mortality. *Science, 260*(5114), 1666—1667. https://doi.org/10.1126/science.8503016

Vaupel, J. W., Carey, J. R., Christensen, K., Johnson, T. E., Yashin, A. I., Holm, N. V., Iachine, I. A., Kannisto, V., Khazaeli, A. A., Liedo, P., Longo, V. D., Zeng, Y., Manton, K. G., & Curtsinger, J. W. (1998). Biodemographic trajectories of longevity. *Science, 280*(5365), 855—860. https://doi.org/10.1126/science.280.5365.855

Vaupel, J. W., Manton, K. G., & Stallard, E. (1979). The impact of heterogeneity in individual frailty on the dynamics of mortality. *Demography, 16*(3), 439—454.

CHAPTER FIFTY

2003—2005: Breakdown in correlations between stress resistance and aging

The standard paradigm

A characteristic feature of nonevolutionary theories for the physiology of aging is the supposition of durable correlations between physiological functions and improved adult survival. Such theories presuppose that if, for example, free-radical damage underlies aging, then the more such free radicals are eliminated physiologically the slower aging will be. By contrast, evolutionary theories of aging make no such assumptions. Instead, in the context of antagonistic pleiotropy, a common expectation of evolutionary biologists is that too much of a physiological investment in one function will lead to countervailing deficiencies in another function. At very extreme imbalances of physiological allocation, evolutionary thinking leads to the expectation that overall life-history should deteriorate.

The conceptual breakthrough

The physiological machinery that was used to test for nonlinear correlations between mechanism and aging was the role of stress resistance in increasing longevity. Phelan et al. (2003) compared a variety of long-evolved Drosophila stocks which had been kept free of inbreeding. At low to moderate levels of increased resistance to acute desiccation or starvation, stress resistance was positively associated with longevity across these populations. But at much higher levels of stress resistance produced by lab selection in the stocks created by Rose et al. (1992), longevity either stabilized or fell.

Archer et al. (2003) used trajectories of direct and indirect responses to selection directly on stress resistance to show such reversals of correlations. For example, over the first 11 generations of selection for resistance to desiccation, longevity progressively increased. But from generations 26—37, longevity fell as desiccation resistance increased. For starvation resistance,

Conceptual Breakthroughs in The Evolutionary Biology of Aging
ISBN: 978-0-12-821545-6
https://doi.org/10.1016/B978-0-12-821545-6.00019-4

average longevity increased by almost 20 days as selection raised starvation resistance by a bit more than 50 h. But at levels of starvation resistance increases greater than 100 h, longevity fell by 6–8 days. Over many populations and a variety of selection procedures, the overall pattern of the physiological relationship between stress resistance and longevity was one of peak longevities with moderate increases in starvation resistance. Thus, too much of a physiological enhancement can be deleterious for aging.

The work performed by Phelan et al. (2003) and Archer et al. (2003) takes on a deeper meaning by connecting the range of stress resistances with their underlying physiological mechanisms. From such studies as Djawdan et al. (1996) for caloric stores and Nghiem et al. (2000) for the desiccation resistance of long-lived flies, we can take away the fact that moderate storage of water, fat, and carbohydrates are good predictors of longevity. Thus moderate selection for stress resistance improves longevity, while extreme increases in stress resistance violate this pattern.

Still further uncoupling of stress resistance and longevity was observed in a study by Rose et al. (2005). The centerpiece of that article was a doctoral dissertation study by H.B. Passananti, which assembled a starting population made up of a four-way cross between long-standing, somewhat inbred, fruit fly stocks. From this starting population, 10 replicate populations were derived, five with early reproduction and five with late reproduction, similarly to the protocols of Rose (1984). After 22 generations, the late-reproduced populations showed significant increases in average longevity compared to the early reproduced populations, as found by Rose (1984) and Luckinbill et al. (1984), for example. But starvation resistance did *not* increase correspondingly, unlike the pattern observed previously from the research of Service et al. (1985) to the research of Phelan et al. (2003). Furthermore, early fecundity did not decrease in the late-reproduced stock; though this may have occurred because all 10 of these experimental populations were adapting to the overall laboratory conditions.

Impact: 7

It is always tempting for biologists to assume that there are simple and consistent relationships among the characters that they study, relationships that are additive or monotonic. But decades of research on the evolutionary physiology of aging in Drosophila has instead revealed greater and greater complexities. Even with a common ancestral population, correlations between physiological and life-history characters can break down at extreme values of such characters. And among different evolutionary lineages, different physiological mechanisms may underlie the evolution of large changes in patterns of aging. The physiology of aging, like the physiology of life itself, is complicated, even at the level of whole organisms and their major organs.

References and further reading

Archer, M. A., Phelan, J. P., Beckman, K. A., & Rose, M. R. (2003). Breakdown in correlations during laboratory evolution. II. Selection on stress resistance in Drosophila populations. *Evolution*, *57*(3), 536—543. https://doi.org/10.1111/j.0014-3820.2003.tb01545.x

Djawdan, M., Sugiyama, T. T., Schlaeger, L. K., Bradley, T. J., & Rose, M. R. (1996). Metabolic aspects of the trade-off between fecundity and longevity in *Drosophila melanogaster*. *Physiological Zoology*, *69*(5), 1176—1195.

Luckinbill, L. S., Arking, R., Clare, M. J., Cirocco, W. C., & Buck, S. A. (1984). Selection for delayed senescence in *Drosophila melanogaster*. *Evolution*, *38*(5), 996—1003. https://doi.org/10.2307/2408433. JSTOR.

Nghiem, D., Gibbs, A. G., Rose, M. R., & Bradley, T. J. (2000). Postponed aging and desiccation resistance in *Drosophila melanogaster*. *Experimental Gerontology*, *35*(8), 957—969. https://doi.org/10.1016/S0531-5565(00)00163-7

Phelan, J. P., Archer, M. A., Beckman, K. A., Chippindale, A. K., Nusbaum, T. J., & Rose, M. R. (2003). Breakdown in correlations during laboratory evolution. I. Comparative analyses of Drosophila populations. *Evolution*, *57*(3), 527—535. https://doi.org/10.1111/j.0014-3820.2003.tb01544.x

Rose, M. R. (1984). Laboratory evolution of postponed senescence in *Drosophila melanogaster*. *Evolution*, *38*(5), 1004—1010. https://doi.org/10.2307/2408434. JSTOR.

Rose, M. R., Passananti, H. B., Chippindale, A. K., Phelan, J. P., Matos, M., Teotónio, H., & Mueller, L. D. (2005). The effects of evolution are local: Evidence from experimental evolution in Drosophila. *Integrative and Comparative Biology*, *45*(3), 486—491. https://doi.org/10.1093/icb/45.3.486

Rose, M. R., Vu, L. N., Park, S. U., & Graves, J. L. (1992). Selection on stress resistance increases longevity in *Drosophila melanogaster*. *Experimental Gerontology*, *27*(2), 241—250. https://doi.org/10.1016/0531-5565(92)90048-5

Service, P. M., & Rose, M. R. (1985). Genetic covariation among life-history components: The effect of novel environments. *Evolution*, *39*(4), 943—945. https://doi.org/10.2307/2408694. JSTOR.

Service, P. M., Hutchinson, E. W., MacKinley, M. D., & Rose, M. R. (1985). Resistance to environmental stress in *Drosophila melanogaster* selected for postponed senescence. *Physiological Zoology*, *58*(4), 380—389.

CHAPTER FIFTY ONE

2007−11: Development of demographic models that separate aging from dying

The standard paradigm

Conventional reductionist and evolutionary thought up to the mid-1990s envisioned aging proceeding all the way to death, without disconti-nuity. Exponential increases in adult mortality conforming to Gompertz equations were believed to be the best way of modeling the weakening of natural selection to the point of inevitable death. This elegant view of the relationship between aging and death began to fall apart in the 1990s, as we have seen. But it would unravel further in the 2000s. Evolu-tionary biologists and others looked more deeply into the possibility of life after aging, and one of the issues that would emerge was the process of dying itself.

The idea of dying as a process distinct from that of aging is implicit in medical practice, but even in that context, they are often confused. Much of internal medicine concerns metabolic disorders that steadily in-crease in incidence and severity with adult age: diabetes, obesity, cardio-vascular disease, and so on. Most general practitioners deal with biological aging in treating their adult patients, since these are all aging-associated diseases.

There is a specific phase in the progression of aging-associated diseases when they switch from being a cause of limitation or debilitation, to shifting toward pervasive impairment in conjunction with a greatly increased risk of imminent death. Examples of this transition are focal cancers starting to spread metastatically and heart attacks leading to congestive heart failure. Af-ter this transition occurs, a more acute process that we might call "dying" seems to take over. But the general contours of this process were not taken into account in most gerontological research before 2005 (We note in pass-ing that the idea of a "dying phase" was proposed by Clarke and Maynard Smith [1961]).

Conceptual Breakthroughs in The Evolutionary Biology of Aging
ISBN: 978-0-12-821545-6
https://doi.org/10.1016/B978-0-12-821545-6.00057-1

The conceptual breakthrough

By 1996, evolutionary theory was reformulated in terms of three life phases for iteroparous organisms. These phases are development, aging, and late life, in that order. In the first or "development" life phase, the force of natural selection is at its strongest, fostering the survival of reproductively fit offspring to contribute to the next generation, purging most early-onset deleterious genes in the process. The following life phase of "aging" is dominated by the decline in the force of natural selection (Hamilton, 1966), commonly associated with age-specific declines in mortality and fecundity. At advanced ages, the pioneering work of Carey et al. (1992) and Curtsinger et al. (1992) revealed late-life plateaus in age-specific mortality, while plateaus were later discovered for female fecundity (Rauser et al., 2006) and male virility (Shahrestani et al., 2012) as well, as discussed in Chapters 38 and 49.

Statistical analysis of the fecundity data of Rauser et al. (2006) revealed the existence of a new and distinct phase of dying prior to the day of death in Drosophila females, which Mueller et al. (2007) characterized quantitatively. This fourth dying phase was termed "the death spiral." Evidence for the existence of death spirals and their qualitative distinctness from aging has since accumulated from multiple studies (Curtsinger, 2015; Mueller et al., 2009, 2011, 2016; Rogina et al., 2007).

To explore this new life phase, Mueller et al. (2007) used the findings and data from Rauser et al. (2006) to develop a new evolutionary model to explain the distinction between dying and aging. This model, called the "Evolutionary Heterogeneity Model of Fecundity," is distinct from the earlier lifespan heterogeneity models of Vapuel et al. (1979). The term "heterogeneity" in the Mueller et al. (2007) model refers to a hypothesized binary state of females, either in the death spiral or not. The aim of the Evolutionary Heterogeneity Model (EHF) is to characterize the difference in egg-laying rates for female Drosophila during the aging phase before and after they enter the death spiral. The assumption is that, in the transition from the aging phase to late life, fecundity should decline at a linear rate until the "fecundity breakday (fbd)" after which fecundity remains constant (Mueller et al., 2007), as outlined in the following expression:

$$f(t) = \begin{cases} c_1 + c_2 t, \text{ if } t \leq fbd \\ c_1 + c_2 fbd, \text{ if } t > fbd \end{cases} \tag{51.1}$$

It is presumed that the decline in fecundity becomes more rapid once a female has entered the death spiral. Mueller et al. (2007) designates the duration of the death spiral as a window of size w, from the start of the death spiral until eventual death (at age d), where the rate of fecundity decline during the death spiral $\tilde{f}(t)$ is given by the following:

$$\tilde{f}(t) = f(d - w) + f(d - w)c_3(w + t - d) \qquad (51.2)$$

This expression is represented visually in Fig. 51.1 which plots the decline in age-specific female fecundity using two slopes, depending on whether the female is in the death spiral or not at age t^*. If the female is in the death spiral at t^*, it will continue linearly until death with the slope pattern f(t^*) c_3. At age fbd, the "fecundity breakday" is reached in which female fecundity plateaus. The window of days between the start of the spiral and eventual death can vary by population but is presumed to be of a fixed length.

Using fecundity data from Rauser et al. (2006), it was possible to estimate the parameters of Eq. (51.1) using standard nonlinear regression techniques (Mueller et al., 2007) to estimate fecundity break-days from observations of

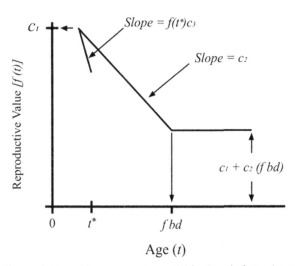

$$\text{Age } (t)$$

Figure 51.1 The evolutionary heterogeneity model for female fecundity in Drosophila, incorporating declining fecundity during aging [slope c_2], fecundity during its late-life plateau [$c_1 + c_2(fbd)$], and the decline in fecundity during the dying phase [slope c_3]. *Inspired from From Mueller, L. D., Rauser, C. L., & Rose, M. R. (2007). An evolutionary heterogeneity model of late-life fecundity in Drosophila.* Biogerontology, 8(2), 147–161. *https://doi.org/10.1007/s10522-006-9042-x*

lifetime age-specific fecundity patterns in 2828 females. When models of this kind are fitted to the average fecundity values plotted in Fig. 49.2, they fit the observed data much better than the simpler aging-to-late-life models shown in that Figure.

Impact: 9

The partitioning of dying individuals from aging individuals who are not yet dying provides greater analytical precision for studies of aging, which previously conflated aging and dying. Subsequent work on individual aging-associated characters showed that models incorporating dying fit observed data far better than models without dying (e.g., Shahrestani et al., 2012), even when the causes of entry into the dying phase were unknown.

From the standpoint of refining the analysis of declining functional characters in gerontology, a four-phase demographic analysis should be used to partition the process of aging from late life and dying, in order to avoid bioinformatic confusion. In the medical context, where the transition to the dying phase is more readily inferred, the distinction between treatments that mitigate aging and treatments that rescue patients from the process of dying could be practically useful.

Finally, we propose that the physiological process of dying may be one source for the plausibility of the nonevolutionary physiological theories of aging. The dramatic disintegration of function, and thus health, which occurs in the dying phase has been seen as the natural conclusion to a physiological process of aging. But now we know that the end of aging may occur as a result of three different life-cycle events: (1) acute trauma, (2) the beginning of a protracted dying process, or (3) the beginning of late life. As commonly occurs with scientific progress, deeper understanding leads to an appreciation of complexities that were previously misperceived or entirely overlooked.

References and further reading

Carey, J. R., Liedo, P., Orozco, D., & Vaupel, J. W. (1992). Slowing of mortality rates at older ages in large medfly cohorts. *Science, 258*(5081), 457–461 (JSTOR).

Clarke, J. M., & Maynard Smith, J. (1961). Two phases of aging in *Drosophila subobscura*. *The Journal of Experimental Biology, 38*, 679–684.

Curtsinger, J. W. (2015). The retired fly: Detecting life history transition in individual *Drosophila melanogaster* females. *The Journals of Gerontology Series A: Biological Sciences and Medical Sciences, 70*(12), 1455–1460. https://doi.org/10.1093/gerona/glv122

Curtsinger, J. W., Fukui, H. H., Townsend, D. R., & Vaupel, J. W. (1992). Demography of genotypes: Failure of the limited life-span paradigm in *Drosophila melanogaster*. *Science, 258*(5081), 461–463. https://doi.org/10.1126/science.1411541

Hamilton, W. D. (1966). The moulding of senescence by natural selection. *Journal of Theoretical Biology, 12*(1), 12—45. https://doi.org/10.1016/0022-5193(66)90184-6

Mueller, L. D., Rauser, C. L., & Rose, M. R. (2007). An evolutionary heterogeneity model of late-life fecundity in Drosophila. *Biogerontology, 8*(2), 147—161. https://doi.org/10.1007/s10522-006-9042-x

Mueller, L. D., Rauser, C. L., & Rose, M. R. (2011). *Does aging stop?* Oxford University Press. https://doi.org/10.1093/acprof:oso/9780199754229.001.0001

Mueller, L. D., Shahrestani, P., & Rauser, C. L. (2009). Predicting death in female Drosophila. *Experimental Gerontology, 44*(12), 766—772. https://doi.org/10.1016/j.exger.2009.09.001

Mueller, L. D., Shahrestani, P., Rauser, C. L., & Rose, M. R. (2016). The death spiral: Predicting death in Drosophila cohorts. *Biogerontology, 17*(5—6), 805—816. https://doi.org/10.1007/s10522-016-9639-7

Rauser, C. L., Tierney, J. J., Gunion, S. M., Covarrubias, G. M., Mueller, L. D., & Rose, M. R. (2006). Evolution of late-life fecundity in *Drosophila melanogaster. Journal of Evolutionary Biology, 19*(1), 289—301. https://doi.org/10.1111/j.1420-9101.2005.00966.x

Rogina, B., & Helfand, S. L. (2013). Indy mutations and Drosophila longevity. *Frontiers in Genetics, 4*, 47. https://doi.org/10.3389/fgene.2013.0004. Published 2013 April 8.

Rogina, B., Wolverton, T., Bross, T. G., Chen, K., Müller, H.-G., & Carey, J. R. (2007). Distinct biological epochs in the reproductive life of female *Drosophila melanogaster. Mechanisms of Ageing and Development, 128*(9), 477—485. https://doi.org/10.1016/j.mad.2007.06.004

Shahrestani, P., Quach, J., Mueller, L. D., & Rose, M. R. (2012). Paradoxical physiological transitions from aging to late life in *Drosophila. Rejuvenation Research, 15*(1), 49—58. https://doi.org/10.1089/rej.2011.1201

Vaupel, J. W., Manton, K. G., & Stallard, E. (1979). The impact of heterogeneity in individual frailty on the dynamics of mortality. *Demography, 16*(3), 439—454.

CHAPTER FIFTY TWO

2010: Studying the evolutionary origins of aging in bacteria

The standard paradigm

Until the early 21st century, the evolutionary origins of aging were left largely unaddressed. However, the prevailing assumption was that the phenomena of aging arose among multicellular eukaryotes.

The evolutionary logic underpinning this assumption was that a unicellular organism that underwent some type of cumulative aging process would transmit that cumulative aging to each and every daughter cell produced by symmetrical fission. With each generation further accumulating such deleterious effects, the lineage would eventually die out. All fissile unicellular organisms, it was supposed, therefore evolved so as to preclude aging entirely, producing only fully functional daughter cells. For example, it was inferred that such unicellular species would effectively sustain cellular metabolism, dispose of metabolic waste products, and keep all other types of intrinsic damage at a tolerable level.

The conceptual breakthrough

These notions were challenged by early reports suggesting that observable aging arose in bacteria (Ackermann et al., 2003; Shapiro et al., 2002; Stewart et al., 2005), with unequal distribution of subcellular structures leading to heterogeneity between the daughter cells produced by fissile reproduction. Specifically, it was found that bacterial reproduction could lead to so much asymmetry that fission could yield a "mother cell" which largely retains older predivision subcellular structures, together with a "daughter cell" that inherits newly synthesized structures. Consequently, the mother cells could suffer progressively worsened function, while producing rejuvenated progeny (Ackermann, 2007). [It should be noted that asymmetrical division and "reproductive aging" have been well-characterized in yeast of the species *Saccharomyces cerevisiae* (Egilmez and Jazwinski, 1989; Mortimer & Johnston, 1959).]

Conceptual Breakthroughs in The Evolutionary Biology of Aging
ISBN: 978-0-12-821545-6
https://doi.org/10.1016/B978-0-12-821545-6.00022-4

A fully developed analysis of this phenomenon was produced by Chao (2010). He formally modeled the evolutionary origins of this sort of asymmetrical fission arising from the asymmetrical partitioning of damage, waste, etc. The chief focus of the theory developed by Chao (2010) is "damage load", a concept broadly analogous to the common population-genetic hypothesis of "mutational load". Damage load arises at the dynamical equilibrium achieved between the generation of cellular damage and phenotypic selection against overly damaged cells, in this model.

With damage load, the problem of accumulating waste products and the like can favor the asymmetrical partition of such wastes. The evolution of such asymmetrical partitioning results in the introduction of the first forms of aging. However, evolution does not always favor the evolution of such asymmetrical reproduction. Under sufficiently benign conditions, and thus less damage or waste accumulation prior to reproduction, Chao (2010) found that unicellular species can sustain symmetrical division and perpetual biological immortality.

Impact: 8

The conceptual breakthrough of Chao (2010) led to a substantial exploration of the evolutionary origins of asymmetrical fission, both in theory and in experimental tests of such theory (Chao et al., 2016; Proenca et al., 2018, 2019; Rang et al., 2011, 2012, 2018). Although much work is still needed to consolidate our understanding of damage, repair, and rejuvenation in fissile unicellular organisms, the work of Chao and colleagues has provided good foundations for research on the evolutionary origins of aging.

We must emphasize, however, that the phrase "damage load" misleadingly suggests that the work of Chao and colleagues supports physiological theories which suppose that aging is nothing more or less than damage to macromolecules. But their theory does not require that particular assumption. For example, the differential partitioning of metabolic wastes or damaged organelles among the products of fission is just two of many types of physiological partitioning that could cause the evolutionary origin of aging in their theory. Furthermore, the Chao theory is explicitly an evolutionary one, not one which assumes that there is an inherent "entropic" necessity to the accumulation of any type of physiological damage, physiological waste products, and so on.

References and further reading

Ackermann, M., Chao, L., Bergstrom, C. T., & Doebeli, M. (2007). On the evolutionary origin of aging. *Aging Cell, 6*(2), 235−244. https://doi.org/10.1111/j.1474-9726.2007.00281.x

Ackermann, M., Stearns, S. C., & Jenal, U. (2003). Senescence in a bacterium with asymmetric division. *Science, 300*(5627), 1920. https://doi.org/10.1126/science.1083532

Chao, L. (2010). A model for damage load and its implications for the evolution of bacterial aging. *PLoS Genetics, 6*(8), e1001076. https://doi.org/10.1371/journal.pgen.1001076

Chao, L., Rang, C. U., Proenca, A. M., & Chao, J. U. (2016). Asymmetrical damage partitioning in bacteria: A model for the evolution of stochasticity, determinism, and genetic assimilation. *PLOS Computational Biology, 12*(1), e1004700. https://doi.org/10.1371/journal.pcbi.1004700

Egilmez, N. K., Chen, J. B., & Jazwinski, S. M. (1989). Specific alterations in transcript prevalence during the yeast life span. *The Journal of Biological Chemistry, 264*(24), 14312−14317.

Lindner, A. B., Madden, R., Demarez, A., Stewart, E. J., & Taddei, F. (2008). Asymmetric segregation of protein aggregates is associated with cellular aging and rejuvenation. *Proceedings of the National Academy of Sciences, 105*(8), 3076−3081. https://doi.org/10.1073/pnas.0708931105

Mortimer, R. K., & Johnston, J. R. (1959). Life span of individual yeast cells. *Nature, 183*(4677), 1751−1752. https://doi.org/10.1038/1831751a0

Proenca, A. M., Rang, C. U., Buetz, C., Shi, C., & Chao, L. (2018). Age structure landscapes emerge from the equilibrium between aging and rejuvenation in bacterial populations. *Nature Communications, 9*(1), 3722. https://doi.org/10.1038/s41467-018-06154-9

Proenca, A. M., Rang, C. U., Qiu, A., Shi, C., & Chao, L. (2019). Cell aging preserves cellular immortality in the presence of lethal levels of damage. *PLOS Biology, 17*(5), e3000266. https://doi.org/10.1371/journal.pbio.3000266

Rang, C. U., Peng, A. Y., & Chao, L. (2011). Temporal dynamics of bacterial aging and rejuvenation. *Current Biology, 21*(21), 1813−1816. https://doi.org/10.1016/j.cub.2011.09.018

Rang, C. U., Peng, A. Y., Poon, A. F., & Chao, L. (2012). Ageing in *Escherichia coli* requires damage by an extrinsic agent. *Microbiology, 158*(6), 1553−1559. https://doi.org/10.1099/mic.0.057240-0

Rang, C. U., Proenca, A., Buetz, C., Shi, C., & Chao, L. (2018). Minicells as a damage disposal mechanism in *Escherichia coli*. *MSphere, 3*(5). https://doi.org/10.1128/mSphere.00428-18

Shapiro, L., McAdams, H. H., & Losick, R. (2002). Generating and exploiting polarity in bacteria. *Science, 298*(5600), 1942−1946. https://doi.org/10.1126/science.1072163

Stewart, E. J., Madden, R., Paul, G., & Taddei, F. (2005). Aging and death in an organism that reproduces by morphologically symmetric division. *PLoS Biology, 3*(2), e45. https://doi.org/10.1371/journal.pbio.0030045

CHAPTER FIFTY THREE

2010: Genome-wide sequencing of evolved aging reveals many sites

The standard paradigm

Before 2010, the conventional gerontological paradigm was that there were a few key "longevity" genes that underlie aging, genes that could be discovered using mutagenesis and related molecular genetic techniques, like the construction of transgenics. The predominant population genetics theory, the "neoclassical model" (cf. Lewontin, 1974), has emphasized the role of rare, large-effect, substitutions in phenotypic change, with most standing genetic variation expected to be neutral or deleterious. This evolutionary process of rare beneficial substitutions has been bolstered by the study of adaptation in asexual populations (e.g., Barrick & Lenski, 2009; Lenski et al., 1991). The neoclassical view of adaptation focused on patterns of selective sweeps; in sexual populations, such selective sweeps are expected to locally deplete genetic variation in chromosomal regions flanking the locations of the alleles favored in these sweeps (vid. Maynard Smith & Haigh, 1974; Burke, 2012). Thus neoclassical evolutionary genetics bolstered the views held by conventional gerontologists, with their emphasis on a small number of important "longevity genes."

But within evolutionary biology, an alternative "balance" framework emphasized the importance of selectively maintained genetic polymorphism in sexual populations (e.g., Dobzhansky, 1970; Lewontin, 1974), assuming that such genetic variation is functionally important and therefore used in adaptation. A formal experiment testing the merits of either model in relation to genetic variance, and the nature of "sweeps", in sexual organisms had not yet been accomplished. The research of the Rose laboratory (e.g., Rose et al., 2004) was predicated on antagonistic pleiotropy sustaining abundant functional genetic variation for aging-related characters, because such antagonistic pleiotropy is one of the balancing selection mechanisms that can

Conceptual Breakthroughs in The Evolutionary Biology of Aging
ISBN: 978-0-12-821545-6
https://doi.org/10.1016/B978-0-12-821545-6.00050-9

maintain selectable genetic variation (e.g., Rose, 1982, 1985). But the existence of such abundant, selectable, genetic variation had not been demonstrated directly at the genomic level before 2010.

The conceptual breakthrough

Burke et al. (2010) pioneered genome-wide "evolve-and-rese-
quence" (also "E&R") work with sexual populations. Their study established that there is extensive genome-wide divergence associated with the
adaptation of 10 populations of *Drosophila melanogaster*, five "A-types" and
five "C-types" subjected to two distinct selection regimes. "A" populations
are maintained with nine to 10-day, discrete-generation, life cycles, while
"C" populations have 28-day life cycles (vid. Chippindale et al., 1997).
This resulted in strong differentiation with respect to patterns of aging,
with A adult cohorts living only 60% as long as C cohorts at the time of
the study (Burke et al., 2010).

Due to budgetary limitations, replicates were pooled within treatment
for genome-wide sequencing. These pooled sequencing data were then
compared for differentiation among the Single-Nucleotide Polymorphisms
(SNP) frequencies across the major chromosome arms. Although all 10 populations had undergone hundreds of generations of selection, genetic diversity was *not* purged near sites undergoing selection, contrary to the
neoclassical model. Instead, the genome-wide results strongly supported
the neobalance model, possibly thanks to the tendency of antagonistic pleiotropy to sustain functional genetic variation. These populations were later
sequenced with greater replication, as shown in Fig. 53.1, where patterns of
SNP divergence are displayed genome-wide.

Impact: 9

What began with Burke et al. (2010) continued with multiple subsequent publications (Remolina et al., 2012; Carnes et al., 2015; Graves et al.,
2017; Fabian et al., 2018), as evolutionary biologists have continued to
sequence Drosophila genomes from populations that have evolved increased
lifespans. Over the course of this research, the level of replication and statistical power have increased substantially. This work collectively reveals that
many sites are involved in the short-term evolution of aging, the obvious
implication being that still more sites in the genome underlie the evolution
of aging in general, beyond the limited scope of experimental evolution.

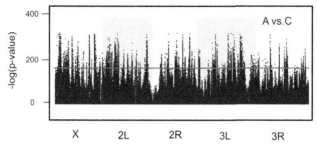

Figure 53.1 A versus C Cochran-Mantel-Haenszel test outputs comparing SNP frequencies between contrasting selection regimes. Results were plotted as $-\log(P\text{-}$values) across all major chromosome arms. The red lines represent the significance threshold from permutation tests. *Data from Graves, J. L., Hertweck, K. L., Phillips, M. A., Han, M. V., Cabral, L. G., Barter, T. T., Greer, L. F., Burke, M. K., Mueller, L. D., & Rose, M. R. (2017). Genomics of parallel experimental evolution in Drosophila. Molecular Biology and Evolution, msw282. https://doi.org/10.1093/molbev/msw282*

An important question that has come up repeatedly in the analysis of the results obtained from such genome-wide sequencing projects is the extent to which the "longevity" mutants of Drosophila or nematodes or mice are represented among the genomic sites involved in the experimental evolution of aging. The answer is that "longevity" mutants are identified less by genome-wide analysis of the experimental evolution of aging than would be expected by chance (Graves et al., 2017; Rose & Burke, 2011). Thus, mutants that sometimes have large effects on aging, relative to their inbred progenitor strains (vid. Khazaeli et al., 2005), fail to reveal the genome-wide foundations for the evolution of aging in outbred populations.

There are two possible explanations for this which are not mutually exclusive. First, nematode aging-mutant research, covered in Chapter 48, appears to involve mutations that are actively selected against, at least under some conditions. Such large-effect "longevity mutants" therefore may be unlikely to segregate in outbred populations, whether in the wild or in laboratories, because natural selection eliminates them.

A second possible explanation for the inadequate representation of "longevity genes" in the genome-wide analysis is epistatic effects involving inbred lines, discussed in Chapter 23. Inbred lines are poor representatives of the outbred populations from which they are derived. Mutations that allow particular inbred lines to live longer in particular environmental conditions were shown to be unreliable in their effects on lifespan by Khazaeli et al. (2005), as discussed in Chapter 47. In that same study, experimentally evolved differences in the lifespans of outbred populations were shown to be readily reproducible, even with very different assay methods. For the purposes of evolutionary research on aging, there is little apparent value in the use of longevity mutants. Whether such mutants are always selected against or only "work" for specific inbred stocks, they do not appear to be reliable guides to the genetics involved in the evolution of aging. In any case, genome-wide research with outbred experimentally evolved stocks can entirely circumvent their use, as we show in this and subsequent chapters.

Whether or not conventional gerontologists wish to continue to conduct research with genetic variants that have little to do with aging in outbred populations is, in large part, up to them. If their interest is in outbred human populations, then research based on "longevity genes" will be of little biomedical value. On the other hand, if they want to help decipher the vagaries of the inbred rodent strains that have been of such obsessive interest to the American National Institutes of Health, then they should continue to work on the problem of irreproducibility in mutant strains derived from inbred lines of Drosophila and the like (vid. Khazaeli et al., 2005).

References and further reading

Barrick, J. E., & Lenski, R. E. (2009). Genome-wide mutational diversity in an evolving population of *Escherichia coli*. *Cold Spring Harbor Symposia on Quantitative Biology, 74*(0), 119—129. https://doi.org/10.1101/sqb.2009.74.018

Burke, M. K. (2012). How does adaptation sweep through the genome? Insights from long-term selection experiments. *Proceedings of the Royal Society B: Biological Sciences, 279*(1749), 5029—5038. https://doi.org/10.1098/rspb.2012.0799

Burke, M. K., Dunham, J. P., Shahrestani, P., Thornton, K. R., Rose, M. R., & Long, A. D. (2010). Genome-wide analysis of a long-term evolution experiment with Drosophila. *Nature, 467*(7315), 587—590. https://doi.org/10.1038/nature09352

Burke, M. K., & Long, A. D. (2012). What paths do advantageous alleles take during short-term evolutionary change? *Molecular Ecology, 21*(20), 4913—4916. https://doi.org/10.1111/j.1365-294X.2012.05745.x

Carnes, M. U., Campbell, T., Huang, W., Butler, D. G., Carbone, M. A., Duncan, L. H., Harbajan, S. V., King, E. M., Peterson, K. R., Weitzel, A., Zhou, S., & Mackay, T. F. C. (2015). The genomic basis of postponed senescence in *Drosophila melanogaster*. *PLOS ONE, 10*(9), e0138569. https://doi.org/10.1371/journal.pone.0138569

Chippindale, A. K., Alipaz, J. A., Chen, H.-W., & Rose, M. R. (1997). Experimental evolution of accelerated development in *Drosophila*. 1. Developmental speed and larval survival. *Evolution, 51*(5), 1536—1551. https://doi.org/10.1111/j.1558-5646.1997.tb01477.x

Dobzhansky, T. (1970). *Genetics of the evolutionary process* (4th ed.). Columbia University Press.

Fabian, D. K., Garschall, K., Klepsatel, P., Santos-Matos, G., Sucena, É., Kapun, M., Lemaitre, B., Schlötterer, C., Arking, R., & Flatt, T. (2018). Evolution of longevity improves immunity in *Drosophila. Evolution Letters, 2*(6), 567—579. https://doi.org/10.1002/evl3.89

Graves, J. L., Hertweck, K. L., Phillips, M. A., Han, M. V., Cabral, L. G., Barter, T. T., Greer, L. F., Burke, M. K., Mueller, L. D., & Rose, M. R. (2017). Genomics of parallel experimental evolution in *Drosophila. Molecular Biology and Evolution.* https://doi.org/10.1093/molbev/msw282. msw282.

Khazaeli, A., Vanvoorhies, W., & Curtsinger, J. (2005). The relationship between life span and adult body size is highly strain-specific in. *Experimental Gerontology, 40*(5), 377—385. https://doi.org/10.1016/j.exger.2005.02.004

Lenski, R. E., Rose, M. R., Simpson, S. C., & Tadler, S. C. (1991). Long-term experimental evolution in *Escherichia coli.* I. Adaptation and divergence during 2,000 generations. *The American Naturalist, 138*(6), 1315—1341. https://doi.org/10.1086/285289

Lewontin, R. C. (1974). *The genetic basis of evolutionary change.* Columbia University Press.

Maynard Smith, J., & Haigh, J. (1974). The hitch-hiking effect of a favourable gene. *Genetical Research, 23*(1), 23—35. https://doi.org/10.1017/S0016672300014634

Remolina, S. C., Chang, P. L., Leips, J., Nuzhdin, S. V., & Hughes, K. A. (2012). Genomic basis of aging and life-history evolution in *Drosophila melangaster.* Genomics of life-history evolution. *Evolution, 66*(11), 3390—3403. https://doi.org/10.1111/j.1558-5646.2012.01710.x

Rose, M. R. (1982). Antagonistic pleiotropy, dominance, and genetic variation. *Heredity, 48*(1), 63—78. https://doi.org/10.1038/hdy.1982.7

Rose, M. R. (1985). Life history evolution with antagonistic pleiotropy and overlapping generations. *Theoretical Population Biology, 28*(3), 342—358. https://doi.org/10.1016/0040-5809(85)90034-6

Rose, M. R., & Burke, M. K. (2011). Genomic Croesus: Experimental evolutionary genetics of Drosophila aging. *Experimental Gerontology, 46*(5), 397—403. https://doi.org/10.1016/j.exger.2010.08.025

Rose, M. R., Passananti, H. B., & Matos, M. (2004). *Methuselah flies: A case study in the evolution of aging.* World Scientific Publishing. https://doi.org/10.1142/5457

CHAPTER FIFTY FOUR

2011—19: Evolutionary transcriptomics also reveal complex physiology of aging

The standard paradigm

Research using genome-wide sequencing, primarily in Drosophila populations (vid. Burke et al., 2010; Graves et al., 2017), provided concrete evidence that many genomic sites are involved in the short-term evolution of aging. The conventional reductionist paradigm for aging could be rescued, however, if different patterns of aging were underlain by a few central transcriptional pathways which corresponded to pathways identified from studies of large-effect mutants (e.g., *methuselah* and *Indy*). The efficacy of using techniques that involve knocking out candidate genes (e.g., Bray et al., 2016) to assess the impact of such genes on longevity through a limited number of specific pathways hinges on this principle.

The conceptual breakthrough

The aim of early transcriptome studies, like those reviewed by Sarup et al. (2011), was to use gene expression to probe transcriptional mechanisms underlying adult survival with and without stress. It was unclear at that time, however, just how universal candidate genes would be across different studies. To test this, their study used data from six different transcriptomic studies of *D. melanogaster*. What they observed was a high degree of overlap of candidate genes from populations with similar ancestry, but *not* with different ancestries. The findings of Sarup et al. (2011) highlight the degree to which genetic drift and possibly small differences in selection procedures can obscure reliable inference of transcriptomic differentiation across multiple studies.

One of the first genome-wide expression rate studies to target aging specifically was the work of Remolina et al. (2012) which, unlike predecessors that tracked a single population undergoing changing life phases, instead compared separate populations with divergent aging patterns. After 50

Conceptual Breakthroughs in The Evolutionary Biology of Aging
ISBN: 978-0-12-821545-6
https://doi.org/10.1016/B978-0-12-821545-6.00003-0

generations of divergent selection between treatments, whole genome sequencing and transcriptome profiling were used to identify 156 genes deemed to differentiate in response to selection on patterns of aging. Remolina et al. (2012) pioneered the use of multi-level omic characterization of experimental evolution to elucidate the molecular basis of aging. Integral to their work was the use of highly differentiated populations derived from a common founding population, in order to distinguish the effects of selection from stochastic processes.

This kind of experiment was taken to an extreme with the transcriptomic analyses of Barter et al. (2019), which incorporated tenfold population replication for each of two selection treatments, "A" versus "C", with sharply differing aging patterns. The A-types begin the aging process approximately 2 weeks prior to their C-type counterparts, as shown in Fig. 54.1 (Burke et al., 2016). These stark phenotypic differences were underlain by the strong genome-wide differences shown in Fig. 54.2 (Graves et al., 2017).

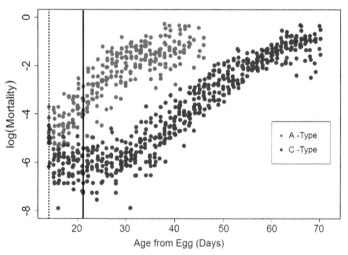

Figure 54.1 Adult age-specific mortality (females) from 10 A-type and 10 C-type selected populations. Mortality rates were normalized using logarithmic transformation, each point presents mortality over 2-day intervals. The dotted line represents the sampling timepoint for day 14 versus the solid bar for the collection sample at day 21. *Data from Burke, M. K., Barter, T. T., Cabral, L. G., Kezos, J. N., Phillips, M. A., Rutledge, G. A., Phung, K. H., Chen, R. H., Nguyen, H. D., Mueller, L. D., & Rose, M. R. (2016). Rapid divergence and convergence of life-history in experimentally evolved* Drosophila melanogaster. Evolution, 70(9), 2085–2098. https://doi.org/10.1111/evo.13006

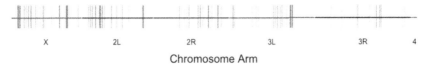

Figure 54.2 Sites of genomic differentiation between 10 A-type and 10 C-type selected populations. Using poolseq on Drosophila females, each line represents the location of a differentiated SNP region (50 kb) within the Drosophila genome. Each chromosome arm is color coded. *Data from Graves, J.L., Hertweck, K.L., Phillips, M.A., Han, M.V., Cabral, L.G., Barter, T.T., Greer, L.F., Burke, M.K., Mueller, L.D., & Rose, M.R. (2017). Genomics of Parallel Experimental Evolution in Drosophila. Molecular Biology and Evolution, msw282. https://doi.org/10.1093/molbev/msw282.*

Barter et al. (2019) characterized the transcriptomes of these populations to determine the gene-expression differences between cohorts that were either aging or not aging. This was accomplished by sampling both sets of populations in parallel on day 14 and day 21 from eggs, ages at which the A-types had already begun to age, while the C-types remained in a pre-aging phase featuring the stable low-mortality rates shown in Fig. 54.1. The transcriptomic analysis identified 906 transcripts that were differentially expressed in the 10 A populations compared to the 10 C populations. These differentiated transcripts were dispersed widely across all chromosomes, as shown in Fig. 54.3. This extensive and complex pattern of gene-expression differences that were reproducibly associated with the presence versus absence of aging matches the complexity of the genome-wide differentiation found by Burke et al. (2010) and Graves et al. (2017), once again suggesting that a simple reductionist approach to the molecular machinery of the evolution of aging is deeply flawed.

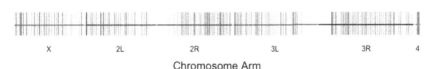

Figure 54.3 Sites of transcriptomic differentiation between 10 A-type and 10 C-type selected populations. Each line represents the location of a differentiated transcript from the Drosophila genome. Each chromosome arm is color coded with the same designated colors as in Fig. 54.2. *Data from Barter, T. T., Greenspan, Z. S., Phillips, M. A., Mueller, L. D., Rose, M. R., & Ranz, J. M. (2019). Drosophila transcriptomics with and without ageing. Biogerontology, 20(5), 699—710. https://doi.org/10.1007/s10522-019-09823-4*

Impact: 9

Starting with Sarup et al. (2011), Remolina et al. (2012), and culminating with Barter et al. (2019), the last decade of research has shown that the response to selection on Drosophila aging and related characters is underlain by hundreds of transcriptomic differences across the genome. As with the genome-sequencing research, the more replication that was employed in these transcriptomic studies, the more transcriptional differences they revealed.

Finally, it should be noted once again that transcriptomic research does not support a central evolutionary role for the pathways identified from reductionist research based on large-effect "longevity" Drosophila mutants discussed in Chapter 47. The transcripts of "longevity genes" are not particularly represented in the results of powerful transcriptome-wide research. Evidently, this echoes the findings of the previous Chapter 53. The conventional cell-molecular reductionist paradigm for the analysis of aging in outbreeding species still appears to be scientifically moribund, when tested precisely using genome-wide and transcriptome-wide data from outbred populations. Shifting from genome-wide research to transcriptome-wide research still does not rescue reductionist gerontology, with no sign of the mechanistic simplicity that it requires.

References and further reading

Barter, T. T., Greenspan, Z. S., Phillips, M. A., Mueller, L. D., Rose, M. R., & Ranz, J. M. (2019). Drosophila transcriptomics with and without ageing. *Biogerontology*, *20*(5), 699—710. https://doi.org/10.1007/s10522-019-09823-4

Bray, N. L., Pimentel, H., Melsted, P., & Pachter, L. (2016). Erratum: Near-optimal prob-abilistic RNA-seq quantification. *Nature Biotechnology*, *34*(8), 888. https://doi.org/10.1038/nbt0816-888d

Burke, M. K., Barter, T. T., Cabral, L. G., Kezos, J. N., Phillips, M. A., Rutledge, G. A., Phung, K. H., Chen, R. H., Nguyen, H. D., Mueller, L. D., & Rose, M. R. (2016). Rapid divergence and convergence of life-history in experimentally evolved *Drosophila melanogaster*. *Evolution*, *70*(9), 2085—2098. https://doi.org/10.1111/evo.13006

Burke, M. K., Dunham, J. P., Shahrestani, P., Thornton, K. R., Rose, M. R., & Long, A. D. (2010). Genome-wide analysis of a long-term evolution experiment with Drosophila. *Nature*, *467*(7315), 587—590. https://doi.org/10.1038/nature09352

Graves, J. L., Hertweck, K. L., Phillips, M. A., Han, M. V., Cabral, L. G., Barter, T. T., Greer, L. F., Burke, M. K., Mueller, L. D., & Rose, M. R. (2017). Genomics of parallel experimental evolution in *Drosophila*. *Molecular Biology and Evolution*. https://doi.org/10.1093/molbev/msw282. msw282.

Remolina, S. C., Chang, P. L., Leips, J., Nuzhdin, S. V., & Hughes, K. A. (2012). Genomic basis of aging and life-history evolution in *Drosophila melangaster*. Genomics of life-history evolution. *Evolution*, *66*(11), 3390—3403. https://doi.org/10.1111/j.1558-5646.2012.01710.x

Sarup, P., Sørensen, J. G., Kristensen, T. N., Hoffmann, A. A., Loeschcke, V., Paige, K. N., & Sørensen, P. (2011). Candidate genes detected in transcriptome studies are strongly dependent on genetic background. *PLoS ONE, 6*(1), e15644. https://doi.org/10.1371/journal.pone.0015644

CHAPTER FIFTY FIVE

2012: Late life is physiologically different from aging

The standard paradigm

Given the existence of late life as a life-history phase after aging, the question arises as to whether its physiology is different from that of aging. From the standpoint of evolutionary biology, late life in adult organisms is a unique period in which the forces of natural selection no longer differentiate among age classes. When there is an absence of direct natural selection, prior experimental studies have demonstrated that the evolutionary stabilization of mortality and female fecundity depends on the specific timing of the cessation of natural selection (Rauser et al., 2006; Rose et al., 2002). That is, late life is a contingent product of the forces of natural selection, not some general problem of "reliability" or "lifelong heterogeneity" independent of evolution (vid. Mueller et al., 2011).

This poses the important question, however, whether or not physiological characters plateau the way survival and fertility do, or if they exhibit more complex patterns in late life. Among the possible patterns could be (a) continuation of aging trends, (b) cessation of the physiological deterioration characteristic of aging, and (c) reversal of aging trends, as shown in Fig. 55.1. If, however, all the physiological declines observed during aging continued unremittingly after the transition to demographic late life, then applying Hamiltonian theory to the evolutionary physiology of life phases would be rendered relatively difficult.

The conceptual breakthrough

Shahrestani, Quach, et al. (2012) studied the trajectories of age-specific organismal physiological characters as Drosophila with different lifespans transitioned from aging to late life. Prior work on late life (Mueller et al., 2009; Shahrestani, Tran, et al., 2012; Rauser et al., 2006) had established that in late life fitness characteristics like mortality, fecundity, and

Conceptual Breakthroughs in The Evolutionary Biology of Aging
ISBN: 978-0-12-821545-6
https://doi.org/10.1016/B978-0-12-821545-6.00059-5

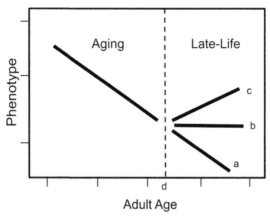

Figure 55.1 Potential patterns for the physiological transitions from "aging" to "late life": (A) Continuation of aging trends, (B) cessation of the deterioration characteristic of aging, and (C) reversal of aging trends.

virility all plateau. The last point was established by Shahrestani, Tran, et al. (2012), which showed that virility in male *D. melanogaster* also plateaued, as predicted by the Hamiltonian theory of late life (Fig. 55.2), further

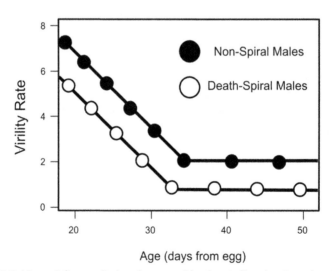

Figure 55.2 Mean virility predictions between "death-spiral" males, 1 week or less from death, compared to "nonspiral" males living 1 week or more. Simulations predicted lower intercepts and earlier break-days compared to those not in the death spiral. Both groups demonstrate similar rates of vitality decline, though overall lower virility in "death-spiral" males. *Data from Shahrestani, P., Tran, X., & Mueller, L.D. (2012b). Patterns of male fitness conform to predictions of evolutionary models of late life: Late-life male fitness. Journal of Evolutionary Biology, 25(6), 1060–1065. https://doi.org/10. 1111/j.1420-9101.2012.02492.x.*

corroborating the evolutionary theory of aging and demonstrating that fertility plateaus were not sex-specific.

Shahrestani, Quach, et al. (2012) sought to test whether this same pattern held for other physiological characters during the transition between deteriorating fitness characters versus their stabilization. Using six replicate populations (B_{1-5}/IV), 57,000+ *D. melanogaster* flies held in cages as adults were assayed, both before and after the onset of late life, for desiccation resistance, time spent in motion, negative geotaxis, and starvation resistance (Shahrestani, Quach, et al., 2012). Adult mortality rates were used to delineate the aging phase from the late phase when mortality rates plateaued.

In almost all cases, the transition from aging to late life was marked by a change in the trajectory of age-specific physiology. In some cases, such as starvation resistance in cohorts handled in cages, late-life physiology featured a cessation or slowing of the physiological deterioration characteristic of aging. But other characters, such as climbing, showed no such slowing of deterioration. This went against the general notion that late life should always be a period of physiological stabilization.

These results were extended in the work of Shahrestani et al. (2016), which explored the effects of evolutionary history on the timing of late life transitions. Presumably, if aging and late life are distinct from one another due to differing strengths in the force of natural selection, then the transition between aging and late life should shift depending on the age at which selection pressures plateau. Shahrestani et al. (2016) compared physiological transitions in "A-type" accelerated populations with those of longer-lived "C-type" 28-day cycle counterparts with respect to mortality, desiccation resistance, time spent in motion, negative geotaxis, and starvation resistance. As expected from evolutionary theories of late life, the A-types demonstrated not only earlier mortality plateaus but also earlier transitions from aging-phase declines to the more heterogeneous patterns previously observed for physiological characters during late life, compared to their longer-lived C-type ancestors (See Fig. 55.3).

Impact: 7

Shahrestani, Quach, et al. (2012), Shahrestani et al. (2016) demonstrated that aging and late life are often distinguishable from one another in the physiology of *Drosophila melanogaster*. Given that late life is not directly subject to the forces of natural selection, its functional features cannot be expected to show the consistent pattern of deterioration exhibited during

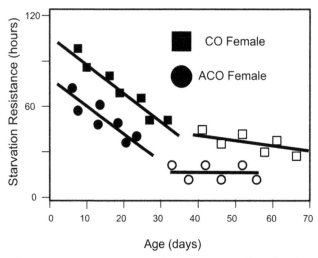

Age (days)

Figure 55.3 Starvation resistance patterns between ACO and CO females. Filled symbols represent age-specific mean values prior to the mortality breakday and open symbols are the means postbreakday *Adaptation of data from Shahrestani, P., Wilson, J. B., Mueller, L. D., & Rose, M. R. (2016). Patterns of physiological decline due to age and selection in* Drosophila melanogaster. *Evolution, 70(11), 2550–2561. https://doi.org/10. 1111/evo.13065*

aging when both fitness characteristics and functional physiology predictably deteriorate as the forces of natural selection fall. In a sense, late life is akin to the development phase, when natural selection does not consistently distinguish among specific age classes, producing more varied temporal patterns than those of aging.

However, the interpretation of these findings is complicated by the potential involvement of genotype-by-interaction. In the study of Shahrestani, Quach, et al. (2012), flies that had never been maintained in cages during their evolutionary histories were nonetheless held in cages prior to physiological assays. In that case, the late–life trajectories of age-specific characters included some cases of continued decline. But in the study of Shahrestani et al. (2016), the Drosophila cohorts that were held in cages came from populations that had long evolved in laboratory cages. This second study showed general stabilization of physiological functions during late life. For the time being, then, we are not sure if physiological stabilization in late life is likely to occur when populations are maintained in environments to which they have long adapted. But it is a testable hypothesis.

References and further reading

Mueller, L. D., Rauser, C. L., & Rose, M. R. (2011). *Does aging stop?* Oxford: Oxford University Press.

Mueller, L. D., Shahrestani, P., & Rauser, C. L. (2009). Predicting death in female Drosophila. *Experimental Gerontology, 44*(12), 766—772. https://doi.org/10.1016/j.exger.2009.09.001

Rauser, C. L., Tierney, J. J., Gunion, S. M., Covarrubias, G. M., Mueller, L. D., & Rose, M. R. (2006). Evolution of late-life fecundity in *Drosophila melanogaster. Journal of Evolutionary Biology, 19*(1), 289—301. https://doi.org/10.1111/j.1420-9101.2005.00966.x

Rose, M. R., Drapeau, M. D., Yazdi, P. G., Shah, K. H., Moise, D. B., Thakar, R. R., Rauser, C. L., & Mueller, L. D. (2002). Evolution of late-life mortality in Drosophila melanogaster. *Evolution, 56*(10), 1982—1991. https://doi.org/10.1111/j.0014-3820.2002.tb00124.x

Shahrestani, P., Quach, J., Mueller, L. D., & Rose, M. R. (2012a). Paradoxical physiological transitions from aging to late life in *Drosophila. Rejuvenation Research, 15*(1), 49—58. https://doi.org/10.1089/rej.2011.1201

Shahrestani, P., Tran, X., & Mueller, L. D. (2012b). Patterns of male fitness conform to predictions of evolutionary models of late life: Late-life male fitness. *Journal of Evolutionary Biology, 25*(6), 1060—1065. https://doi.org/10.1111/j.1420-9101.2012.02492.x

Shahrestani, P., Wilson, J. B., Mueller, L. D., & Rose, M. R. (2016). Patterns of physiological decline due to age and selection in *Drosophila melanogaster. Evolution, 70*(11), 2550—2561. https://doi.org/10.1111/evo.13065

CHAPTER FIFTY SIX

2014: Genomic studies of centenarians have low scientific power

The standard paradigm

A hallmark paradigm of conventional gerontological research is to compare individual humans or animals who attain great chronological age with the general population from which they come. A recurring design flaw with such association studies is their notable lack of statistical power, given poor sample sizes of long-lived individuals (Tan et al., 2008). At present, the average life-expectancy of humans ranges from ~80 to 85 years of age, with only 1%–2% reaching near centenarian status. Twin and other studies suggest that approximately 20%–30% of genetic variation correlates with heritable longevity (e.g., Herskind et al., 1996). The ambition of human genome-wide association studies (GWAS) of longevity in centenarians is to scan for genes that are significantly differentiated between centenarians and the general population, in order to determine which genes in centenarians permit their relatively greater longevity, if indeed any.

The conceptual breakthrough

Burke et al. (2014) used four synthetic *D. melanogaster* populations, each generated by crossing eight inbred lines from the Drosophila Synthetic Population Resource (DSPR) (King, Macdonald et al., 2012; King, Merkes et al., 2012). Each of these four synthetic populations was used to generate a cohort of adult flies. Samples were collected at random from among the early adult members of each cohort. In addition, those flies remaining after 98% of the cohort had died were sampled as experimental analogs of centenarians among human populations.

Burke et al. (2014) compared the sequences of genomes of the randomly chosen cohort members with the genomes of those that had survived long enough to achieve "centenarian" status. Their analysis sought to identify

Conceptual Breakthroughs in The Evolutionary Biology of Aging
ISBN: 978-0-12-821545-6
https://doi.org/10.1016/B978-0-12-821545-6.00037-6

the "haplotype" regions of the genome which were significantly differentiated between randomly sampled individuals and the "centenarians."

This research was an improvement over traditional centenarian studies, in that the genetic origins of each cohort were well-defined, the sampling regime was unbiased, and there was genome-wide sequencing. Despite that, Burke et al. found no consistent results among the four populations. Even within individual cohorts, they found little genomic differentiation between long-living flies and the fly cohort in general.

Impact: 3

The work of Burke et al. (2014) can be contrasted with the previous work by Burke et al. (2010), described in Chapter 53. Despite the far greater amount of genome-wide sequencing performed by Burke et al. (2014), compared to Burke et al. (2010), consistent results were not obtained (The study of Graves et al. (2017) which was larger but similar in design to that of Burke et al. (2010), also described in Chapter 53, yielded extensive and highly reproducible results). This contrast suggests that the experimental design of Burke et al. (2014) was appreciably misconceived.

There are two possible sources for the failure of this study. The first possibility is that centenarian studies generally feature too little genomic differentiation between phenotypically undifferentiated individuals and the centenarians chosen on the basis of their phenotypes. The second possibility is that the use of hybrids of inbred lines, or inbred lines themselves, is a bad design choice. In Chapter 25, we discussed an early study of Rose (1984) in which it was shown that moderate levels of inbreeding distorted genetic correlations systematically. Furthermore, genetic research on moderate inbreeding in samples of *D. subobscura* has shown that less inbreeding than that used to create the ancestral lines crossed for the study of Burke et al. (2014) produces strikingly chaotic stochastic effects (Santos et al., 2013).

Finally, it should be added that the studies of Burke et al. (2010), Remolina et al. (2012), and Graves et al. (2017) employed replicated selection lines, in which deterministic effects can be reproducible. By contrast, the Burke et al. (2014) project involved differentiation initially produced by the stochastic effects of inbreeding. Given the statistical challenges facing any genome-wide research project, relying on random genetic differentiation versus selectively produced genetic differentiation will be a bad experimental design choice, unless the topic of interest is the impact of small effective population sizes itself. In Chapter 57, we will return to these issues in the broader context of studying the full range of evolving species, across the diversity of mating systems.

References and further reading

Burke, M. K., Dunham, J. P., Shahrestani, P., Thornton, K. R., Rose, M. R., & Long, A. D. (2010). Genome-wide analysis of a long-term evolution experiment with Drosophila. *Nature, 467*(7315), 587—590. https://doi.org/10.1038/nature09352

Burke, M. K., King, E. G., Shahrestani, P., Rose, M. R., & Long, A. D. (2014). Genome-wide association study of extreme longevity in Drosophila melanogaster. *Genome Biology and Evolution, 6*(1), 1—11. https://doi.org/10.1093/gbe/evt180

Graves, J. L., Hertweck, K. L., Phillips, M. A., Han, M. V., Cabral, L. G., Barter, T. T., Greer, L. F., Burke, M. K., Mueller, L. D., & Rose, M. R. (2017). Genomics of parallel experimental evolution in Drosophila. *Molecular Biology and Evolution.* https://doi.org/10.1093/molbev/msw282. msw282.

Herskind, A. M., McGue, M., Holm, N. V., Sørensen, T. I. A., Harvald, B., & Vaupel, J. W. (1996). The heritability of human longevity: A population-based study of 2872 Danish twin pairs born 1870—1900. *Human Genetics, 97*(3), 319—323. https://doi.org/10.1007/BF02185763

King, E. G., Macdonald, S. J., & Long, A. D. (2012a). Properties and power of the Drosophila synthetic population Resource for the routine dissection of complex traits. *Genetics, 191*(3), 935—949. https://doi.org/10.1534/genetics.112.138537

King, E. G., Merkes, C. M., McNeil, C. L., Hoofer, S. R., Sen, S., Broman, K. W., Long, A. D., & Macdonald, S. J. (2012b). Genetic dissection of a model complex trait using the Drosophila Synthetic Population Resource. *Genome Research, 22*(8), 1558—1566. https://doi.org/10.1101/gr.134031.111

Remolina, S. C., Chang, P. L., Leips, J., Nuzhdin, S. V., & Hughes, K. A. (2012). Genomic basis of aging and life-history evolution in Drosophila melangaster. Genomics of life-history evolution. *Evolution, 66*(11), 3390—3403. https://doi.org/10.1111/j.1558-5646.2012.01710.x

Rose, M. R. (1984). Genetic covariation in Drosophila life history: Untangling the data. *The American Naturalist, 123*(4), 565—569. https://doi.org/10.1086/284222

Santos, J., Pascual, M., Simões, P., Fragata, I., Rose, M. R., & Matos, M. (2013). Fast evolutionary genetic differentiation during experimental colonizations. *Journal of Genetics, 92*(2), 183—194. https://doi.org/10.1007/s12041-013-0239-x

Tan, Q., Zhao, J. H., Zhang, D., Kruse, T. A., & Christensen, K. (2008). Power for genetic association study of human longevity using the case-control design. *American Journal of Epidemiology, 168*(8), 890—896. https://doi.org/10.1093/aje/kwn205

CHAPTER FIFTY SEVEN

2015: Evolutionary genetic effects produce two evolutionary biologies of aging

The standard paradigm

A common view among reductionist gerontologists and some evolutionary biologists is that the genetic and physiological mechanisms of aging among asexual species are likely to be similar to those of outbreeding species. Specifically, a commonly assumed evolutionary genetic mechanism for selective change is selective "sweeps" in which new mutations that benefit fitness rise from low frequencies to near-fixation (vid. Burke, 2012). Such substitutions are expected to occur among "longevity assurance" genes, leading to the evolution of varied patterns of aging according to the effectiveness of such longevity assurance genes.

Many experiments exploring the evolutionary genetics of naturally outbred eukaryotic populations have adopted the *"neoclassical"* paradigm over the last 40 years, a paradigm discussed in greater detail in Chapter 53. The neoclassical view expects that both adaptation and its associated patterns of aging depend on rare beneficial mutations that sweep to fixation.

If this neoclassical model for evolutionary genetics theory were generally correct, there would be no practical distinction between asexual and sexual populations in the study of the evolution of their aging. Among other things, studies of both adaptation and aging could study selective sweeps that would identify key genetic sites affecting fitness and aging (vid. Burke, 2012).

The conceptual breakthrough

Rose et al. (2015) brought attention to the important differences in the evolution of aging between species that do not have frequent recombination from those that do have frequent recombination. Species that do

Conceptual Breakthroughs in The Evolutionary Biology of Aging
ISBN: 978-0-12-821545-6
https://doi.org/10.1016/B978-0-12-821545-6.00045-5
243

not have frequent recombination evolve their adaptations, and thus their aging, chiefly because of selective sweeps of alleles with moderate to large effects on fitness (e.g., Barrick et al., 2009; Lenski et al., 1991; Tenaillon et al., 2012). A side-effect of such selective sweeps in asexual species is a tendency for genetic variation to be largely purged, across the entire genome.

By contrast, species that *do* have frequent recombination, on the other hand, demonstrably do not usually evolve by such selective sweeps (vid. Phillips et al., 2016). Even when they do have selective sweeps, on rare occasions, the depression of genetic variation will be local to genomic sites immediately flanking the site of the selective sweep (Burke, 2012). Thus, outbred species will evolve patterns of aging that depend on the tuning of genetic variation at many sites in the genome, where allelic differences at these sites will often be of individually small effect (vid. Graves et al., 2017; see Fig. 57.1).

Rose et al. (2015) argued that this clear disparity implies the use of different experimental methods when studying aging in asexual versus sexual species. In species that rarely undergo recombination, aging is usefully

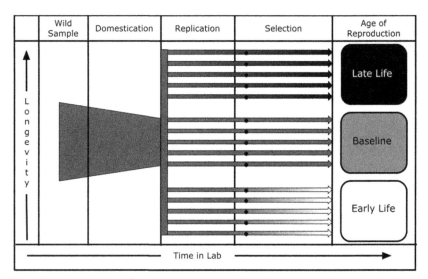

Figure 57.1 Experimental evolution can manipulate the observed phenotypic patterns of longevity by changing the rearing protocol and ultimately altering the Hamiltonian forces of natural selection. Stemming from common wild-type ancestors, each population is differentiated at many sites across the genome, wholly different from what is observed in asexual systems.

examined by studies of mutants and transgenics with moderate to large effects on functional characters, especially components of fitness. By contrast, sexual species will have extensive genetic variation across many sites in the genome, genome-wide variation which allows their aging to be reproducibly shaped by changing the timing of reproduction, as discussed extensively in this book.

An important point that has been raised implicitly in Chapters 25 and 53 is the use of inbred lines, or their hybrids, derived from normally outbred sexual species. Practically speaking, it is hard to study such material using either quantitative genetics (vid. Chapter 25) or genome-wide sequencing (vid. Chapters 53 and 56). It is especially dubious to apply inferences from results obtained with artificially inbred material to the evolution of aging in asexual or outbred sexual species, as was demonstrated in the research discussed in Chapters 53 and 54 (Fig. 57.1).

Impact: 6

Although experimental scientists treat asexual and sexual systems as markedly different in practice, a clear explanation of why the interpretation of results from these two kinds of genetic systems should be distinct is provided by Rose et al. (2015). This is particularly important for aging research, because most who study the genetics of aging do so with an interest in addressing biomedical problems associated with the physiological declines of human aging. For fundamental progress to be made in the field of gerontology, both in understanding and remedying the decline in fitness components with the fall of natural selection, such studies must be done in the light of appropriate evolutionary theory.

Reductionist gerontology has its place in elucidating the genetics and physiology of nonrecombinant asexual organisms, because such species will often evolve through the substitution of beneficial mutants with large effects that can analyzed using conventional Mendelian and molecular genetic manipulations, such as mutagenesis and transgenic manipulations. But reductionist gerontology has limited use in parsing the genetics and physiology of aging in normally outbreeding species. Thus, in the context of biomedical research of value for treating the ailments of human aging, reductionist methods will only rarely be of appreciable utility. Instead, biomedical research based on genome-wide, transcriptome-wide, and metabolomic studies of outbred populations, whether model species or humans themselves, will be of greatest value.

References and further reading

Barrick, J. E., Yu, D. S., Yoon, S. H., Jeong, H., Oh, T. K., Schneider, D., Lenski, R. E., & Kim, J. F. (2009). Genome evolution and adaptation in a long-term experiment with *Escherichia coli*. *Nature, 461*(7268), 1243—1247. https://doi.org/10.1038/nature08480

Burke, M. K. (2012). How does adaptation sweep through the genome? Insights from long-term selection experiments. *Proceedings of the Royal Society B: Biological Sciences, 279*(1749), 5029—5038. https://doi.org/10.1098/rspb.2012.0799

Graves, J. L., Hertweck, K. L., Phillips, M. A., Han, M. V., Cabral, L. G., Barter, T. T., Greer, L. F., Burke, M. K., Mueller, L. D., & Rose, M. R. (2017). Genomics of parallel experimental evolution in *Drosophila*. *Molecular Biology and Evolution*. https://doi.org/ 10.1093/molbev/msw282. msw282.

Lenski, R. E., Rose, M. R., Simpson, S. C., & Tadler, S. C. (1991). Long-Term experimental evolution in *Escherichia coli*. I. Adaptation and divergence during 2,000 generations. *The American Naturalist, 138*(6), 1315—1341. https://doi.org/10.1086/ 285289

Maynard Smith, J., & Haigh, J. (1974). The hitch-hiking effect of a favourable gene. *Genetical Research, 23*(1), 23—35. https://doi.org/10.1017/S0016672300014634

Phillips, M. A., Long, A. D., Greenspan, Z. S., Greer, L. F., Burke, M. K., Villeponteau, B., Matsagas, K. C., Rizza, C. L., Mueller, L. D., & Rose, M. R. (2016). Genome-wide analysis of long-term evolutionary domestication in *Drosophila melanogaster*. *Scientific Reports, 6*(1), 39281. https://doi.org/10.1038/srep39281

Rose, M. R., Cabral, L. G., Philips, M. A., Rutledge, G. A., Phung, K. H., Mueller, L. D., & Greer, L. F. (2015). The great evolutionary divide: Two genomic systems Biologies of aging. In A. I. Yashin, & S. M. Jazwinski (Eds.), *Interdisciplinary topics in gerontology* (Vol. 40, pp. 63—73). S. KARGER AG. https://doi.org/10.1159/000364930

Tenaillon, O., Rodríguez-Verdugo, A., Gaut, R. L., McDonald, P., Bennett, A. F., Long, A. D., & Gaut, B. S. (2012). The molecular diversity of adaptive convergence. *Science, 335*(6067), 457—461. https://doi.org/10.1126/science.1212986

CHAPTER FIFTY EIGHT

2016: Experimental evolution can produce nonaging young adults

The standard paradigm

There are two conventional views as to when aging starts. One is that aging starts with the beginning of each individual organism's life. The other is that aging starts only with the start of reproduction, and thus biological adulthood. The latter is seemingly in keeping with Hamilton's (1966) original result that falling mortality should not start to evolve until the onset of reproduction in a population's evolutionary history. But that interpretation is not correct, as we will now show.

The conceptual breakthrough

Hamiltonian evolutionary theory does not require that the start of a population's reproduction coincide with the onset of biological adulthood, where the latter is defined as the production of gametes and the onset of mating. An unanticipated finding was published by Burke et al. (2016), who measured adult mortality rates in large cohorts obtained from Drosophila populations which had been cultured for hundreds of generations without reproduction until more than 2 weeks after the onset of biological adulthood. When assayed under conditions similar to those that they were subjected to for hundreds of generations, these adult cohorts did not exhibit rising mortality rates for their first $20-25$ days of adult life (Fig. 58.1). They were biological adults, but they were not aging.

The cohorts that exhibited this pattern had been cultured for 350 generations (CO) and 200 generations (nCO), respectively, with only eggs laid from days $26-28$ (relative to the beginning of their lives as eggs) used to start the next generation. In terms of Hamilton's (1966) force of natural selection acting on survival, this culture regime kept selection at full force during all days prior to day 26 for these populations. Despite having reached reproductive maturity as early as day 14, there was no fall in the force of natural selection acting on survival until after day 26.

Conceptual Breakthroughs in The Evolutionary Biology of Aging
ISBN: 978-0-12-821545-6
https://doi.org/10.1016/B978-0-12-821545-6.00062-5
247

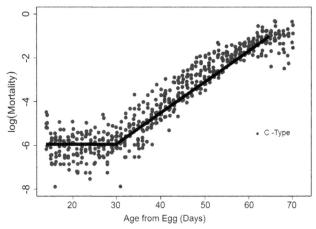

Figure 58.1 Adult age-specific mortality in Drosophila females from cohorts obtained from 10 C-type selected populations. Points represent log-transformed mortality rate per population in each selection treatment. A distinct nonaging plateau is observed between ages 14—28 prior to the forces of natural selection declining, though these flies are already sexually mature adults. *Data from Burke, M. K., Barter, T. T., Cabral, L. G., Kezos, J. N., Phillips, M. A., Rutledge, G. A., Phung, K. H., Chen, R. H., Nguyen, H. D., Mueller, L. D., & Rose, M. R. (2016). Rapid divergence and convergence of life-history in experimentally evolved Drosophila melanogaster. Evolution, 70(9), 2085—2098. https://doi.org/10.1111/ evo.13006*

Prior studies working with the CO populations had *not* revealed this absence of aging, however (e.g., Rose et al., 2002). Burke et al. (2016) attributed this disparity of results to differences in the specific handling methods used in the two assays. The CO cohorts were assayed for mortality rates using vials in the experiments of Rose et al. (2002). These same populations were assayed for mortality using the cages in which they had been cultured for hundreds of generations in the experiments of Burke et al. (2016). Thus, the latter assay regime closely matched the environmental conditions employed during the experimental evolution of the studied populations, while the former assay regime did *not* match those conditions.

This contrast apparently produced an important genotype-by-environment interaction, a common source of artifacts in the study of both experimental evolution (vid. Leroi et al., 1994a, b; Kurapati et al., 2000) and aging (vid. Khazaeli et al., 2005), as discussed repeatedly in this book. Thus, the stable and low mortality rates observed during early adult ages in the study of Burke et al. (2016) are expected to be a more accurate reflection of the evolution of aging among C-type populations than the preservation of aging in early adulthood found by Rose et al. (2002).

Impact: 8

The experimental evolution of nonaging younger adults is a corollary of the evolutionary theory of aging, because these adults were obtained from populations where such young adults had not been permitted to reproduce for hundreds of generations. But this possibility had never been observed before. This finding is a definitive falsification of theories of aging that are not based on Hamilton's (1966) forces of natural selection. Aging is entirely a pattern of age-specific adaptation, *not* a physiological deterioration that inevitably starts with the onset of life *or* the onset of reproduction. As such, the timing of both the onset and the cessation of aging is determined by the ages at which successful reproduction begins and ends in the evolutionary history of a population, since the latter timings define the range of ages over which the forces of natural selection decline (vid. Mueller et al., 2011; Rose et al., 2012).

References and further reading

Burke, M. K., Barter, T. T., Cabral, L. G., Kezos, J. N., Phillips, M. A., Rutledge, G. A., Phung, K. H., Chen, R. H., Nguyen, H. D., Mueller, L. D., & Rose, M. R. (2016). Rapid divergence and convergence of life-history in experimentally evolved *Drosophila melanogaster*. *Evolution, 70*(9), 2085—2098. https://doi.org/10.1111/evo.13006

Hamilton, W. D. (1966). The moulding of senescence by natural selection. *Journal of Theoretical Biology, 12*(1), 12—45. https://doi.org/10.1016/0022-5193(66)90184-6

Khazaeli, A., Vanvoorhies, W., & Curtsinger, J. (2005). The relationship between life span and adult body size is highly strain-specific in *Drosophila melanogaster*. *Experimental Gerontology, 40*(5), 377—385. https://doi.org/10.1016/j.exger.2005.02.004

Kurapati, R., Passananti, H. B., Rose, M. R., & Tower, J. (2000). Increased hsp22 RNA levels in Drosophila lines genetically selected for increased longevity. *The Journals of Gerontology: Series A, 55*(11), B552—B559. https://doi.org/10.1093/gerona/55.11.B552

Leroi, A. M., Chippindale, A. K., & Rose, M. R. (1994). Long-term laboratory evolution of a genetic life-history trade-off in *Drosophila melanogaster*. 1. The role of genotype-by-environment interaction. *Evolution, 48*(4), 1244—1257. https://doi.org/10.2307/2410382. JSTOR.

Leroi, A. M., Chen, W. R., & Rose, M. R. (1994). Long-term laboratory evolution of a genetic life-history trade-off in *Drosophila melanogaster*. 2. Stability of genetic correlations. *Evolution, 48*(4), 1258—1268. https://doi.org/10.1111/j.1558-5646.1994.tb05310.x

Mueller, L. D., Rauser, C. L., & Rose, M. R. (2011). *Does aging stop?* Oxford: Oxford University Press.

Rose, M. R., Drapeau, M. D., Yazdi, P. G., Shah, K. H., Moise, D. B., Thakar, R. R., Rauser, C. L., & Mueller, L. D. (2002). Evolution of late-life mortality in *Drosophila melanogaster*. *Evolution, 56*(10), 1982—1991. https://doi.org/10.1111/j.0014-3820.2002.tb00124.x

Rose, M. R., Flatt, T., Graves, J. L., Greer, L. F., Martinez, D. E., Matos, M., Mueller, L. D., Shmookler Reis, R. J., & Shahrestani, P. (2012). What is aging? In *Frontiers in genetics* (Vol. 3). https://doi.org/10.3389/fgene.2012.00134

CHAPTER FIFTY NINE

2017: The heart is implicated in the evolution of aging

The standard paradigm

A common criticism of evolutionary research on aging is that it is not based on candidate mechanisms, particularly at the physiological level. This critique is fair to that extent that evolutionary biologists working on aging have been more focused on life history characters in relation to genetics rather than exploring specific molecular pathways involved in the aging process.

Over the past 40 years, Drosophila has become a key invertebrate system for studying physiological processes in aging. An early focus of evolutionary physiology research on aging was stress resistance, especially in Drosophila (e.g., Service et al., 1985), as well as locomotor performance (e.g., Graves et al., 1988). That work in turn implicated aspects of lipid and glycogen storage (e.g., Djawdan et al., 1998; Gibbs et al., 1997). However, fine delineation of the specific roles of individual tissues and organs was not achieved in that research.

The conceptual breakthrough

Starting in 2017, it was discovered that the experimental evolution of aging and stress resistance in *Drosophila melanogaster* also depends on heart function (Hardy et al., 2018; Kezos et al., 2019; Shahrestani et al., 2017). Specifically, heart failure under electrical "pacing" (electrical stimulation that speeds up heart-rate) is correlated with patterns of aging: Drosophila populations with more robust hearts live longer (Shahrestani et al., 2017). On the other hand, Drosophila populations that have evolved increased starvation resistance are much fatter, which in turn impinges on heart robustness (Hardy et al., 2018; Kezos et al., 2019). Starvation-resistant and obese Drosophila populations that have been evolved in the laboratory with replication and matched controls are a powerful tool not only for studying the evolution of starvation responses, but also for studying metabolic disorders and related cardiac dysfunction. Hardy et al. observed dilated hearts and

Conceptual Breakthroughs in The Evolutionary Biology of Aging
ISBN: 978-0-12-821545-6
https://doi.org/10.1016/B978-0-12-821545-6.00014-5

reduced contractility in their three evolved obese populations after 65 generations of selection for starvation resistance.

The genomic foundations of such heart differentiation are complex, and they do not implicate the well-known Drosophila "heart mutations," like *tinman* or *opa1* (vid. Bodmer, 1993; Ocorr et al., 2007; Shahrestani et al., 2009). Instead, the genomics of these starvation-resistant but heart-impaired populations involve many differentiated loci, none of which appear to be genetic sites of those classic "heart mutants" (e.g., Kezos et al., 2019). The kinds of heart disease that are common among present-day human populations are unlikely to result from such deleterious alleles of major effect on the heart, the "heart mutants," except in the extremely rare cases of genetic disease, like *opa1* (Shahrestani et al., 2009). It would be impossible for these alleles to rise to high frequencies in outbred populations, because of natural selection acting to remove them, like the longevity mutants discussed in Chapters 23 and 47.

Impact: 7

The evolutionary physiology of aging is starting to build bridges between the evolutionary control of aging and its associated organismal characters, on one hand, and a more mechanistic understanding of how the organs of longer-lived Drosophila are reconfigured when aging evolves in the laboratory. The Drosophila heart is known to have some genetic commonalities with the human heart (vid. Occorr et al., 2007). That makes the evolutionary physiology of the Drosophila heart a promising initial connection between conventional biomedical research on humans and the findings of the evolutionary biology of aging, chiefly in Drosophila. Study of outbred Drosophila populations that have evolved different patterns of heart aging should provide a guide to the physiological and genomic foundations of the cardiovascular diseases which are so common among elderly human patients, at least at the level of overall pattern. It is not yet clear whether the pattern of genetic parallelism that arises with heart mutants in both humans and Drosophila will generalize to the genome-wide sites that influence "normal" heart disease in these two species. Some genetic homologies that connect the two species may indeed be associated with functional parallelism in their effects on heart function. But that is an empirically decidable question, through the comparison of genome-wide and transcriptome-wide studies on outbred populations of the two species.

Similar work on structures like brains, guts, and musculature in Drosophila populations that have evolved different patterns of aging would be welcome. Finally, comparable evolutionary physiology research on other species that have selectively differentiated patterns of aging would be particularly helpful, in order to determine how general the genomic foundations of functional aging are.

References and further reading

Bodmer, R. (1993). The gene tinman is required for specification of the heart and visceral muscles in Drosophila. *Development, 118*(3), 719−729. https://doi.org/10.1242/dev.118.3.719

Bodmer, R. (1995). Heart development in Drosophila and its relationship to vertebrates. *Trends in Cardiovascular Medicine, 5*(1), 21−28. https://doi.org/10.1016/1050-1738(94)00032-Q

Djawdan, M., Chippindale, A. K., Rose, M. R., & Bradley, T. J. (1998). Metabolic reserves and evolved stress resistance in *Drosophila melanogaster*. *Physiological Zoology, 71*(5), 584−594. https://doi.org/10.1086/515963

Gibbs, A. G., Chippindale, A. K., & Rose, M. R. (1997). Physiological mechanisms of evolved desiccation resistance in *Drosophila melanogaster*. *The Journal of Experimental Biology, 200*(Pt 12), 1821−1832.

Graves, J. L., Luckinbill, L. S., & Nichols, A. (1988). Flight duration and wing beat frequency in long- and short-lived *Drosophila melanogaster*. *Journal of Insect Physiology, 34*(11), 1021−1026. https://doi.org/10.1016/0022-1910(88)90201-6

Hardy, C. M., Burke, M. K., Everett, L. J., Han, M. V., Lantz, K. M., & Gibbs, A. G. (2018). Genome-wide analysis of starvation-selected Drosophila melanogaster—a genetic model of obesity. *Molecular Biology and Evolution, 35*(1), 50−65. https://doi.org/10.1093/molbev/msx254

Kezos, J. N., Phillips, M. A., Thomas, M. D., Ewunkem, A. J., Rutledge, G. A., Barter, T. T., Santos, M. A., Wong, B. D., Arnold, K. R., Humphrey, L. A., Yan, A., Nouzille, C., Sanchez, I., Cabral, L. G., Bradley, T. J., Mueller, L. D., Graves, J. L., & Rose, M. R. (2019). Genomics of early cardiac dysfunction and mortality in obese *Drosophila melanogaster*. *Physiological and Biochemical Zoology, 92*(6), 591−611. https://doi.org/10.1086/706099

Ocorr, K., Reeves, N. L., Wessells, R. J., Fink, M., Chen, H.-S. V., Akasaka, T., Yasuda, S., Metzger, J. M., Giles, W., Posakony, J. W., & Bodmer, R. (2007). KCNQ potassium channel mutations cause cardiac arrhythmias in *Drosophila* that mimic the effects of aging. *Proceedings of the National Academy of Sciences, 104*(10), 3943−3948. https://doi.org/10.1073/pnas.0609278104

Service, P. M., Hutchinson, E. W., MacKinley, M. D., & Rose, M. R. (1985). Resistance to environmental stress in *Drosophila melanogaster* selected for postponed senescence. *Physiological Zoology, 58*(4), 380−389.

Shahrestani, P., Burke, M. K., Birse, R., Kezos, J. N., Ocorr, K., Mueller, L. D., Rose, M. R., & Bodmer, R. (2017). Experimental evolution and heart function in *Drosophila*. *Physiological and Biochemical Zoology, 90*(2), 281−293. https://doi.org/10.1086/689288

Shahrestani, P., Leung, H.-T., Le, P. K., Pak, W. L., Tse, S., Ocorr, K., & Huang, T. (2009). Heterozygous mutation of Drosophila Opa1 causes the development of multiple organ abnormalities in an age-dependent and organ-specific manner. *PLoS ONE, 4*(8), e6867. https://doi.org/10.1371/journal.pone.0006867

CHAPTER SIXTY

2020: Evolutionary adaptation to diet and its impact on healthspan

The standard paradigm

Genotype-by-environment interaction has been a long-standing topic in the evolutionary biology of aging together with its associated characters (e.g., Leroi et al., 1994a,b; Service & Rose, 1985) and integral to previous chapters of this book. The kind of life-history trade-offs implicit in the antagonistic pleiotropy theory of aging (e.g., Rose, 1985; Williams, 1957) can appear and disappear with small changes in the assay conditions used in Drosophila experiments (e.g., Leroi et al., 1994a; review in Rose et al., 1996). This dilemma has been acknowledged and demonstrated as a systemic challenge in doing evolutionary research on aging (e.g., Rose, 1991), but its full import was not fully appreciated.

The conceptual breakthrough

Mueller et al. (2011) and Rose et al. (2014) suggested that human aging might exhibit an age-dependent pattern of genotype-by-environment interaction involving diet. Specifically, they proposed that the declining force of natural selection with respect to age might have produced strong and effective human adaptation to an agricultural diet at early ages, but *not* at later ages. These speculations were backed up by simple numerical models.

The real breakthrough was achieved by Rutledge et al. (2020, 2021), working with *Drosophila melanogaster* populations. His study populations had originally been sampled from a long-standing apple orchard population (Ives, 1970), and then subjected to hundreds of generations of culture using banana-molasses medium. Rutledge et al. (2020, 2021) established that fly populations which were long cultivated on the banana medium at early ages had similar or better fertility *at those early ages* when assayed using that same banana medium compared to an apple-based medium. But

Conceptual Breakthroughs in The Evolutionary Biology of Aging
ISBN: 978-0-12-821545-6
https://doi.org/10.1016/B978-0-12-821545-6.00040-6
255

they had measurably superior fertility *at later ages* on apple compared to banana medium, even though they had not been cultured using apple substrate for more than 40 years. This was not an incidental effect of the apple diet. Flies that had been cultured for many generations with banana medium, but using eggs laid only at later ages, showed a durable superiority on banana food across adult ages, compared to their performance on apple food, up until the very last ages of assay (Rutledge et al., 2020). That was because their adaptation to the banana diet continued to much later adult ages (Fig. 60.1).

Impact: 9

The discovery of age-dependent adaptation to an environmental change is significant in two important respects. Firstly, it revealed a generally neglected corollary of Hamilton's forces of natural selection: their fall with adult age leads to an age-dependent wave of adaptation to environmental change. This is a potentially general feature of all evolving populations with age-structure; young individuals should be relatively well-adapted to evolutionarily recent, but not immediate, environmental changes, but not older individuals.

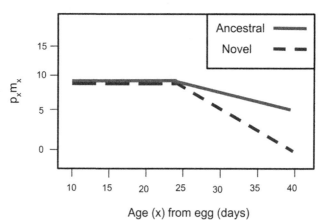

Age (x) from egg (days)

Figure 60.1 Graphical representation of the $p_x m_x$ regression with respect to adult age between ACO flies on ancestral apple diets (*red, solid line*) versus novel banana diets (*blue, dash line*). Inspired by data from Rutledge, G. A., Cabral, L. G., Kuey, B. J., Lee, J. D., Mueller, L. D., & Rose, M. R. (2020). Hamiltonian patterns of age-dependent adaptation to novel environments. PLOS ONE, 15(10), e0240132. https://doi.org/10.1371/journal.pone.0240132

Secondly, since most humans descend from populations that have recently, from an evolutionary perspective, adopted a novel agricultural diet and lifestyle, we should expect that there will be a diet-by-age interaction with respect to optimal human nutrition. That is to say, it is a corollary of the findings of Rutledge et al. (2020, 2021) that humans with longstanding agricultural ancestry should be well-adapted to such foods *only at juvenile and early adult ages*. At later adult ages, humans may have improved health and function on a diet more akin to that of their hunter-gatherer ancestors. Of course, humans without agricultural ancestry are not expected to be adapted to agricultural foods at any age. However, we note that the appropriate delineation of what constitutes appropriate emulations of hunter-gatherer versus agricultural diets among humans is a tricky matter for further research. Most contemporary human diets do not emulate either type of diet. These two implications of the research of Rutledge et al. (2020, 2021) are among the more profound uncovered in evolutionary research on aging, one scientific, the other medical.

References and further reading

Ives, P. T. (1945). The genetic structure of American populations of *Drosophila melanogaster*. *Genetics, 30*(2), 167–196. https://doi.org/10.1093/genetics/30.2.167

Ives, P. T. (1970). Further genetic studies of the south Amherst population of Drosophila melanogaster. *Evolution; International Journal of Organic Evolution, 24*(3), 507–518. https://doi.org/10.1111/j.1558-5646.1970.tb01785.x

Kurapati, R., Passananti, H. B., Rose, M. R., & Tower, J. (2000). Increased hsp22 RNA levels in Drosophila lines genetically selected for increased longevity. *The Journals of Gerontology: Series A, 55*(11), B552–B559. https://doi.org/10.1093/gerona/55.11.B552

Leroi, A. M., Chippindale, A. K., & Rose, M. R. (1994). Long-term laboratory evolution of a genetic life-history trade-off in *Drosophila melanogaster*. 1. The role of genotype-by-environment interaction. *Evolution, 48*(4), 1244–1257. https://doi.org/10.2307/2410382. JSTOR.

Leroi, A. M., Chen, W. R., & Rose, M. R. (1994). Long-term laboratory evolution of a genetic life-history trade-off in Drosophila melanogaster. 2. Stability of genetic correlations. *Evolution, 48*(4), 1258–1268. https://doi.org/10.1111/j.1558-5646.1994.tb05310.x

Mueller, L. D., Rauser, C. L., & Rose, M. R. (2011). *Does aging stop?* Oxford: Oxford University Press.

Rose, M. R. (1985). Life history evolution with antagonistic pleiotropy and overlapping generations. *Theoretical Population Biology, 28*(3), 342–358. https://doi.org/10.1016/0040-5809(85)90034-6

Rose, M. R. (1991). *Evolutionary biology of aging*. Oxford University Press.

Rose, M. R., Cabral, L. G., Philips, M. A., Rutledge, G. A., Phung, K. H., Mueller, L. D., & Greer, L. F. (2014). The great evolutionary divide: Two genomic systems biologies of aging. In A. I. Yashin, & S. M. Jazwinski (Eds.), *Interdisciplinary topics in gerontology* (Vol. 40, pp. 63–73). S. KARGER AG. https://doi.org/10.1159/000364930

Rose, M. R., Nusbaum, T. J., & Chippindale, A. K. (1996). Laboratory evolution: The experimental wonderland and the cheshire cat syndrome. In M. R. Rose, & G. V. Lauder (Eds.), *Adaptation* (pp. 221−241). Academic Press.

Rutledge, G. A., Cabral, L. G., Kuey, B. J., Lee, J. D., Mueller, L. D., & Rose, M. R. (2020). Hamiltonian patterns of age-dependent adaptation to novel environments. *PLOS ONE, 15*(10), e0240132. https://doi.org/10.1371/journal.pone.0240132

Rutledge, G. A., Phang, H. J., Le, M. N., Bui, L., Rose, M. R., Mueller, L. D., & Jafari, M. (2021). Diet and botanical supplementation: Combination therapy for healthspan improvement? *Rejuvenation Research, 24*(5), 331−344. https://doi.org/10.1089/rej.2020.2361

Service, P. M., Hutchinson, E. W., MacKinley, M. D., & Rose, M. R. (1985). Resistance to environmental stress in *Drosophila melanogaster* selected for postponed senescence. *Physiological Zoology, 58*(4), 380−389.

Williams, G. C. (1957). Pleiotropy, natural selection, and the evolution of senescence. *Evolution, 11*(4), 398−411. https://doi.org/10.2307/2406060. JSTOR.

Conclusion

John C. Avise
Department of Ecology and Evolutionary Biology, University of California, Irvine, CA, United States

The scientific value of the evolutionary biology of aging

As this book is a summary of over a century's worth of work, and the material contained in each chapter is not suited to further summarization, we nevertheless feel that there are certain scientific themes that are sustained throughout this volume that bear some emphasis in our conclusion.

Key to the development, if not indeed the success, of the evolutionary biology of aging has been the mathematical analysis of verbal hypotheses and improvements in experimental methods. Like other fields in evolutionary biology as a whole, this field did not lack initial speculations (e.g. Fisher, 1930; Haldane, 1941; Medawar, 1952; Weismann, 1889; Williams, 1957). Fortunately, evolutionary biology has had more than a century of cogent, sometimes trenchant, mathematical formulation and examination of alternative hypotheses. The evolutionary biology of aging has been no exception to this salutary tradition (e.g. Charlesworth, 1980, 2001; Hamilton, 1966; Mueller & Rose, 1996).

But it has usually been challenging for evolutionary biology to test its theories using strong-inference experimentation (cf. Platt, 1964). In this respect, the evolutionary biology of aging has been notably successful among Darwinian fields. Experimental tests of evolutionary theories of aging were not easily achieved at first. Prior to 1980, this experimental literature was thin and often equivocal (e.g. Sokal, 1970; Wattiaux, 1968a, b). In the 1980s, some useful small-scale experiments were done that were more consistent in their durability (e.g. Rose & Charlesworth, 1980; Luckinbill et al., 1984). But by the 1990s, there was a substantial body of strong experimental findings, thanks in large part to work done with *Drosophila* (reviewed in Rose et al., 2004).

A major turn in experimental work on aging came with the ground-breaking 1992 publications demonstrating the existence of a postaging phase of roughly stabilized mortality rates (Carey et al., 1992; Curtsinger et al., 1992). Not only did that discovery have a lasting impact on theories for the evolution of aging (summarized in Mueller et al., 2011), it also transformed experimental research on aging. Instead of using the small cohort sizes characteristic of experimental work in the 1980s, studies with large cohort sizes were used to delimit the initiation, progress, and eventual cessation of aging in the literature of the field starting in the 1990s (e.g. Khazaeli et al., 1998; Rose et al., 2002; Rauser et al., 2006). In effect, the chief focus of the field moved from average lifespan to age-specific mortalities and fecundities throughout adult life. This made strong-inference experimentation much harder, but scientifically still more powerful.

Another point of major scientific progress in the field was the construction of useful bridges from demographic characters, like lifespan, to the underlying physiological machinery of aging. With the advent of selectively postponed aging stocks in the 1980s, it was quickly established that such experimental material provided a powerful window into the physiology of aging (e.g. Graves et al., 1988; Service et al., 1985). It was then shown that manipulating physiology using selection could in turn change patterns of aging (e.g. Archer et al., 2003; Rose et al., 1992). This line of work broke through at the molecular genetic level, especially with the advent of genome-wide sequencing technologies (e.g. Barter et al., 2019; Burke et al., 2010; Graves et al., 2017; Remolina et al., 2012). This helped ground the evolutionary biology of aging on a molecular biological foundation. It also underscored that starting from the premises and tools of evolutionary biology was a powerful experimental strategy for parsing the physiology of aging.

It should be said that almost none of the scientific breakthroughs achieved by the evolutionary biology of aging were rooted in conventional gerontological theories or findings. Indeed, early work by evolutionary biologists on conventional gerontological hypotheses (e.g. Smith, 1958, 1966) would repeatedly falsify such ideas. This pattern of falsification continues to the present day, as illustrated by the discovery of nonaging plateaus during early adulthood by Burke et al. (2016).

To baldly summarize the scientific situation at present, we regard the contest between conventional physiological theories of aging and the evolutionary biology of aging to have reached a critical point similar to that of the contest between the geocentric and the heliocentric theories of astrophysics in the late 17th century. While the geocentric theory is correct in the case of the Moon, as a general model for astrophysics it is unequivocally wrong. From the standpoint of basic science, the theory that aging is the result of universal, entropic, physiological processes has been entirely falsified. When the forces of natural selection do not fall, as in the case of symmetrical fission, aging does not occur. Biological aging is unequivocally *not* a universal side effect of metabolism, across species, or even across adult ages. Nothing about the fundamental controls of aging are readily understood using the conceptual apparatus of conventional reductionist gerontology as it is now configured.

Instead, aging is preeminently the result of age-specific patterns of natural selection, especially their usual weakening with adult age in ovigerous species. This core idea was intuited by Fisher, Haldane, Medawar, and Williams from 1930 to 1957. All the mathematical, comparative, and experimental work that has been performed since on the basis of that insight has built the scientifically potent field of the evolutionary biology of aging. This despite its meager funding relative to the munificent grants and awards afforded to conventional gerontology. Blinkered reductionist faith has proved as impotent in saving conventional gerontology as the Catholic Church's persecution of Copernican astrophysicists proved impotent in the 17th century. It is long past time for the evolutionary biology of aging to be proclaimed as the necessary and sufficient foundation for the field of biological gerontology as a whole.

The medical value of the evolutionary biology of aging

Having made our case for the scientific cogency, validity, and utility of the evolutionary biology of aging, we are left with the question of how its conceptual breakthroughs might be used medically to ameliorate human aging and health in general. Given the ever-present demand for effective solutions to the challenge of aging, the larger public nonetheless remains ignorant with respect to the powerful evolutionary science that could be brought to bear on their chronic age-associated diseases, disorders, and vulnerabilities. No small cause of that ignorance is the obdurate adherence to Aristotelian assumptions about aging as a merely physiological process among mainstream gerontologists, assumptions that we regard as utterly defunct. We have tried to chip away at this profound scientific mistake here. But we are informed by the examples of Pasteur and Koch; the defeat of a prevailing biomedical orthodoxy can take the better part of a century (Rose et al., 2017).

Indeed, we believe that the struggle to scientifically establish and medically employ the microbiological theory of infectious disease from 1840 to 1920 is instructive for the prospects of the evolutionary biology of aging as a foundation for the treatment of aging and its associated diseases (vid. Rose et al., 2016). For much of the 19th century, there was a fairly overt scientific war between the proponents of the long-established, and medically predominant, miasma theory of disease and the microbiological theory of infectious disease (vid. Rose et al., 2017) (The miasma theory supposed that diseases like malaria and cholera were caused by bad-smelling vapors. Thus, the term malaria itself means "bad air"). Many physicians still held on to the miasma theory up to the early 20th century, despite its many experimental refutations at the hands of Pasteur and Koch during the latter part of the 19th century. Most gerontologists still cling to the Aristotelian hypothesis that aging is due to some type of cumulative and normally irreversible physiological process, especially damage to the molecules of cells. This despite the strong-inference refutations of such ideas that we offer throughout this book.

That said, so long as the advances discussed in this volume are overlooked, millions are likely to suffer and die for the failings of mainstream gerontology. The study of aging goes beyond mere curiosity and idle explorations of the inner workings of nature. To understand aging is to understand many of the foundations of human health (Arnold et al., 2020; Sholl, 2021). Though alluring reductionist explanations and treatments for aging have been propagated by popular science figures (e.g. de Grey & Rae, 2007), we hold that most if not all of their claims are unsubstantiated as well as useless. Despite drawing much needed attention toward tangible solutions for aging and its associated maladies, ultimately the field of reductionist gerontology serves primarily as a distraction with respect to the mitigation of human aging.

If aging is to be remedied, biomedical research on aging must confront the complexity of chronic age-associated diseases in humans, from their genomics through their transcriptomics and their metabolomics to their multifold manifestations across organs. We have little faith that verbal stories about simple pathways will be adequate to solving such profound challenges. Large-scale omic and functional analyses are required, work which will yield vast troves of data. Parsing such abundant data in turn requires the use of bioinformatic tools like machine learning to unravel the myriad ways in which the intricacies of aging limit health spans (cf. Arnold et al., 2020; Mueller et al., 2018).

Accepting the evolutionary foundations of aging could eventually mitigate its medical impact. We see three phases over which this mitigation could occur.

First, consider the kind of simple interventions that microbiologists could turn to based on their theory of infection circa 1870, before physicians were generally persuaded that microbiology supplied the best approach to infectious disease. Insightful microbiologists could wash their hands, boil water from dubious sources, and avoid interacting with people showing symptoms of infection. In the same spirit, evolutionary biologists with Eurasian ancestry can take into account results like those of Rutledge, discussed in Chapter 60, to avoid novel processed foods and, at later ages, endeavor to emulate preagricultural diets. For those with ancestors who did not consume agricultural foods prior to 1500, avoidance of agricultural foods at every age might the best option.

In the same vein, it might be advisable to adopt preagricultural lifestyles beyond patterns of food consumption. This could include increased levels of chronic activity, or sustaining closer networks of friends and allies that might be analogous to those which prevailed before agricultural settlement.

Just as microbiologists *circa 1870* could crudely apply their science to avoid infection, despite their relative lack of detailed knowledge, so could evolutionary biologists *now* crudely apply their understanding of the evolution of aging to reduce the impact of environmental novelty on their health at later ages. Naturally, in both cases, there are limits to what individuals could accomplish on their own, in this first phase of the medical application of the evolutionary biology of aging.

Second, consider what occurred after 1900, once physicians, foundations, and even some governments began to accept the microbiological approach to infectious disease. Surgeons began to disinfect their instruments and to strive to sustain antiseptic conditions in their operating theaters. Milk and other foods were pasteurized, to kill bacteria and other infectious microbes. Efforts to purify drinking water became important components of public infrastructure. Crude vaccination programs were started. Much progress was made with reducing deaths due to infection long before microbiologists achieved adequate detail in their scientific characterizations of microbial diversity, mechanisms of infection, or the working of the human immune system.

Similarly, once aging is understood as a complex problem of adaptation declining as a function of age and environment, a variety of broad "public health" reforms might be adopted widely. Older people could be steered away from agricultural foods, from bread to milk. Simple-minded diet supplementation based on defunct gerontological confabulations could be dismissed, from resveratrol to a wide range of antioxidants. Chronic disorders like diabetes and cardiovascular disease could be treated based on the understanding that their "omic" foundations are complex, rather than functions of the accumulation of just one or two substances.

Thirdly, from the 1950s onward, a large microbiological research effort led to the development of (1) effective antibiotics and antivirals, (2) effective vaccines, (3) effective methods for ensuring antiseptic medical procedures, and (4) modern living environments in which drinking water, foods, and household surfaces were kept as pathogen-free as practically possible. The arc of medical progress from 1890 to 1990 was a microbiology-driven revolution in human health, one which radically increased average human lifespan.

Despite the evident complexity of the omic foundations of aging shown here, radical advances in biological technologies could be sufficient to allow eventual manipulation of human aging. At the present time, we cannot predict which future technologies could be successfully deployed to control human aging. But it is certain that they cannot be marshaled based on the parlous work of conventional reductionist gerontology. The omic retuning during the experimental evolution of postponed aging in laboratory species *might* provide a useful guide to technologically advanced medical interventions for aging. For now, however, that is only speculation.

We hope that this installment of the Conceptual Breakthroughs series will not only introduce new students to the evolutionary explanation of aging, but also equip future scientists to take on the pressing biomedical challenges of human aging.

References

Archer, M. A., Phelan, J. P., Beckman, K. A., & Rose, M. R. (2003). Breakdown in correlations during laboratory evolution. II. Selection on stress resistance in Drosophila populations. *Evolution, 57*(3), 536–543. https://doi.org/10.1111/j.0014-3820.2003.tb01545.x

Arnold, K. R., Kezos, J. N., Rutledge, G. A., Barter, T. T., & Rose, M. R. (2020). An evolutionary analysis of health. In J. Sholl, & S. I. S. Rattan (Eds.), *Explaining health across the sciences* (Vol. 12, pp. 13–34). Springer International Publishing. https://doi.org/10.1007/978-3-030-52663-4_2

Barter, T. T., Greenspan, Z. S., Phillips, M. A., Mueller, L. D., Rose, M. R., & Ranz, J. M. (2019). Drosophila transcriptomics with and without ageing. *Biogerontology, 20*(5), 699–710. https://doi.org/10.1007/s10522-019-09823-4

Burke, M. K., Barter, T. T., Cabral, L. G., Kezos, J. N., Phillips, M. A., Rutledge, G. A., Phung, K. H., Chen, R. H., Nguyen, H. D., Mueller, L. D., & Rose, M. R. (2016). Rapid divergence and convergence of life-history in experimentally evolved *Drosophila melanogaster*. *Evolution, 70*(9), 2085–2098. https://doi.org/10.1111/evo.13006

Burke, M. K., Dunham, J. P., Shahrestani, P., Thornton, K. R., Rose, M. R., & Long, A. D. (2010). Genome-wide analysis of a long-term evolution experiment with Drosophila. *Nature, 467*(7315), 587−590. https://doi.org/10.1038/nature09352

Carey, J. R., Liedo, P., Orozco, D., & Vaupel, J. W. (1992). Slowing of mortality rates at older ages in large medfly cohorts. *Science, 258*(5081), 457−461. JSTOR.

Charlesworth, B. (1980). *Evolution in age-structured populations*. Cambridge, U.K: Cambridge University Press.

Charlesworth, B. (2001). Patterns of age-specific means and genetic variances of mortality rates predicted by the mutation-accumulation theory of ageing. *Journal of Theoretical Biology, 210*(1), 47−65. https://doi.org/10.1006/jtbi.2001.2296

Curtsinger, J. W., Fukui, H. H., Townsend, D. R., & Vaupel, J. W. (1992). Demography of genotypes: Failure of the limited life-span paradigm in *Drosophila melanogaster*. *Science, 258*(5081), 461−463. https://doi.org/10.1126/science.1411541

de Grey, A., & Rae, M. (2007). *Ending aging: The rejuvenation breakthroughs that could reverse human aging in our lifetime*. St. Martin's Press.

Fisher, R. A. (1930). *The genetical theory of natural selection*. Clarendon Press. https://doi.org/10.5962/bhl.title.27468

Graves, J. L., Hertweck, K. L., Phillips, M. A., Han, M. V., Cabral, L. G., Barter, T. T., Greer, L. F., Burke, M. K., Mueller, L. D., & Rose, M. R. (2017). Genomics of parallel experimental evolution in Drosophila. *Molecular Biology and Evolution*. https://doi.org/10.1093/molbev/msw282. msw282.

Graves, J. L., Luckinbill, L. S., & Nichols, A. (1988). Flight duration and wing beat frequency in long- and short-lived *Drosophila melanogaster*. *Journal of Insect Physiology, 34*(11), 1021−1026. https://doi.org/10.1016/0022-1910(88)90201-6

Haldane, J. B. S. (1941). *New paths in genetics* (1st ed.). George Allen and Unwin.

Hamilton, W. D. (1966). The moulding of senescence by natural selection. *Journal of Theoretical Biology, 12*(1), 12−45. https://doi.org/10.1016/0022-5193(66)90184-6

Khazaeli, A. A., Pletcher, S. D., & Curtsinger, J. W. (1998). The fractionation experiment: Reducing heterogeneity to investigate age-specific mortality in Drosophila. *Mechanisms of Ageing and Development, 105*(3), 301−317. https://doi.org/10.1016/S0047-6374(98)00102-X

Luckinbill, L. S., Arking, R., Clare, M. J., Cirocco, W. C., & Buck, S. A. (1984). Selection for delayed senescence in *Drosophila melanogaster*. *Evolution, 38*(5), 996−1003. https://doi.org/10.2307/2408433. JSTOR.

Medawar, P. B. (1952). *An unsolved problem of biology*. H.K. Lewis and Co.

Mueller, L. D., Rauser, C. L., & Rose, M. R. (2011). *Does aging stop?* Oxford: Oxford University Press.

Mueller, L. D., Phillips, M. A., Barter, T. T., Greenspan, Z. S., & Rose, M. R. (2018). Genome-wide mapping of gene−phenotype relationships in experimentally evolved populations. *Molecular Biology and Evolution, 35*(8), 2085−2095. https://doi.org/10.1093/molbev/msy113

Mueller, L. D., & Rose, M. R. (1996). Evolutionary theory predicts late-life mortality plateaus. *Proceedings of the National Academy of Sciences of the United States of America, 93*(26), 15249−15253.

Platt, J. R. (1964). Strong inference: Certain systematic methods of scientific thinking may produce much more rapid progress than others. *Science, 146*(3642), 347−353. https://doi.org/10.1126/science.146.3642.347

Rauser, C. L., Tierney, J. J., Gunion, S. M., Covarrubias, G. M., Mueller, L. D., & Rose, M. R. (2006). Evolution of late-life fecundity in *Drosophila melanogaster*. *Journal of Evolutionary Biology, 19*(1), 289−301. https://doi.org/10.1111/j.1420-9101.2005.00966.x

Remolina, S. C., Chang, P. L., Leips, J., Nuzhdin, S. V., & Hughes, K. A. (2012). Genomic basis of aging and life-history evolution in *Drosophila melangaster*. Genomics of life-history evolution. *Evolution, 66*(11), 3390−3403. https://doi.org/10.1111/j.1558-5646.2012.01710.x

Rose, M. R., Cabral, L. G., Kezos, J. N., Barter, T. T., Phillips, M. A., Smith, B. L., & Burnham, T. C. (2016). Four steps toward the control of aging: Following the example of infectious disease. *Biogerontology, 17*(1), 21−31. https://doi.org/10.1007/s10522-015-9588-6

Rose, M., & Charlesworth, B. (1980). A test of evolutionary theories of senescence. *Nature, 287*(5778), 141−142. https://doi.org/10.1038/287141a0

Rose, M. R., Drapeau, M. D., Yazdi, P. G., Shah, K. H., Moise, D. B., Thakar, R. R., Rauser, C. L., & Mueller, L. D. (2002). Evolution of late-life mortality in *Drosophila melanogaster*. *Evolution, 56*(10), 1982−1991. https://doi.org/10.1111/j.0014-3820.2002.tb00124.x

Rose, M. R., Passananti, H. B., & Matos, M. (2004). *Methuselah flies: A case study in the evolution of aging*. World Scientific Publishing. https://doi.org/10.1142/5457

Rose, M. R., Rutledge, G. A., Cabral, L. G., Greer, L. F., Canfield, A. L., & Cervantes, B. G. (2017). Evolution and the future of medicine. In *On human nature* (pp. 695−705). Elsevier. https://doi.org/10.1016/B978-0-12-420190-3.00042-9

Rose, M. R., Vu, L. N., Park, S. U., & Graves, J. L. (1992). Selection on stress resistance increases longevity in *Drosophila melanogaster*. *Experimental Gerontology, 27*(2), 241−250. https://doi.org/10.1016/0531-5565(92)90048-5

Service, P. M., Hutchinson, E. W., MacKinley, M. D., & Rose, M. R. (1985). Resistance to environmental stress in *Drosophila melanogaster* selected for postponed senescence. *Physiological Zoology, 58*(4), 380−389.

Sholl, J. (2021). Can aging research generate a theory of health? *History and Philosophy of the Life Sciences, 43*(2), 45. https://doi.org/10.1007/s40656-021-00402-w

Smith, J. M. (1958). The effects of temperature and of egg-laying on the longevity of *Drosophila Subobscura*. *Journal of Experimental Biology, 35*(4), 832−842.

Smith, J. M. (1966). Theories of aging. In P. L. Krohn (Ed.), *Topics in the Biology of Aging*. New York: Interscience.

Sokal, R. R. (1970). Senescence and genetic load: Evidence from tribolium. *Science (New York, N.Y.), 167*(3926), 1733−1734. https://doi.org/10.1126/science.167.3926.1733

Wattiaux, J. M. (1968a). Parental age effects in *Drosophila pseudoobscura*. *Experimental Gerontology, 3*(1), 55−61. https://doi.org/10.1016/0531-5565(68)90056-9

Wattiaux, J. M. (1968b). Cumulative parental age effects in *Drosophila subobscura*. *Evolution, 22*(2), 406−421. https://doi.org/10.1111/j.1558-5646.1968.tb05908.x

Weismann, A. (1889). *Essays upon heredity and kindred biological problems*. Oxford: Claredon Press.

Williams, G. C. (1957). Pleiotropy, natural selection, and the evolution of senescence. *Evolution, 11*(4), 398−411. https://doi.org/10.2307/2406060. JSTOR.

Glossary

Accelerated aging Moving the onset of senescence towards a younger age in a population.

ACO populations Outbred populations of *D. melanogaster* selected for rapid development, which also exhibit accelerated aging.

Adaptation An attribute that increases Darwinian fitness; or a product of natural selection that increases Darwinian fitness, or the process of natural selection by which Darwinian fitness is increased.

Age-1 A mutant allele in the nematode *Caenorhabditis elegans* that increases lifespan; this allele is also involved in the nematode dauer pathway.

Age-class A group of individuals that share membership in a range of ages, e.g., humans between the ages of 14 and 21 years of age.

Age-distribution Range and frequency of age-groups in a given population.

Age-independent mortality A mortality rate coefficient with a numerical value that is constant across all age groups.

Age-related A biological difference that is statistically significant between two age groups that does not imply dysfunction, specific cause, or a progressive trend continuing to later ages.

Age-structure A population's age-class composition.

Aging The sustained decline in age-specific survival probability or fecundity that takes place after the start of reproduction in some species, regardless of general improvements in feeding, rest, etc.

Agricultural diet A diet that includes agricultural revolution foodstuffs like grains and dairy.

Allele A particular variant form of a gene; e.g., the allele for blue eyes in humans

Amyloid Protein aggregates that bind the dye Congo red and light up as green under polarized light.

Ancestral diet In humans, preagricultural diet that does not include agricultural foodstuffs like grains and dairy. In general, a diet was once characteristic of a population or species, but which was abandoned many generations ago.

Annuals Plants chiefly, but sometimes animals, that have an annual life cycle ended by death after a single season of reproduction.

Antagonistic pleiotropy When allelic variants at a genetic locus have multiple effects, some effects enhancing specific functional characters, like components of fitness, while other effects undermine specific functional characters.

Apoptosis Cell death brought about by internal events, rather than exogenous disruptions like toxins or mechanical damage

Arithmetic mean The arithmetic value obtained by adding up a series of numbers and then dividing the total number count of these numbers.

Artificial selection When experimenters determine the organisms that reproduce based on specific characters, usually measured quantitatively, preventing natural selection as much as possible.

Asexual reproduction Reproduction without genetic exchange with another organism; may involve recombination among the genes of one genome; may involve seeds, buds, branching, or splitting into two symmetrical parts.

B populations Outbred populations of *D. melanogaster* that are maintained using a 2-week life cycle, derived from the Ives (1975) population.

Bacterium   A unicellular microorganism without a true cellular nucleus.

Balancing selection   When natural selection acts to maintain genetic polymorphisms; includes cases with heterozygote superiority and frequency-dependent selection.

Biomarkers of aging   Age-dependent biological characters that are thought to reveal supposed underlying "aging processes"; also, age-dependent biological parameters that predict lifespan, total or remaining.

Cancer   A disease characterized by uncontrolled cellular proliferation.

Carrying capacity   The equilibrium number of individuals that can be supported by an environment.

Centenarian studies   Studies that contrast humans aged greater than 100 years old with those who do not, or who are not expected to, achieve the age of 100 years.

Chromosome   A long string of DNA, sometimes packaged with proteins, usually encoding genes.

Classical model   Evolutionary genetic theory which presumes low levels of segregating Mendelian variation in outbred populations

Clonal senescence, genetic   Falling viability and fertility of clonal organisms when cultured at small population sizes for multiple generations. See also Lansing Effect

Clonal senescence, somatic   Limited replication by somatic cells, in vivo or in vitro, as defined by number of cell doublings. Note that somatic cells that no longer divide may continue to sustain physiological function.

Clone   A group of genetically identical organisms.

CO populations   Outbred populations of *D. melanogaster* selected for moderately postponed aging.

Cohort   A group of individuals of the same species that began life at approximately the same time.

Cole's Paradox   Only a small increase in initial fecundity is required to counterbalance the advantages of repeated reproduction.

Common Ancestor   Pertaining to organisms: an ancestor of two or more contemporaneous individuals. Pertaining to species from which two or more contemporaneous lineages descend.

Conjugation   The exchange of plasmids between bacterial cells by means of pili, projections from 1 cell to another.

Cost of reproduction   A reduction in survival resulting from reproductive activities or structures.

Daf mutations   Mutations that alter the timing or occurrence of "daur" states of developmental arrest in nematodes.

Damage load   Waste products of cellular metabolism in clonal organisms.

Daur   Developmental arrest characterized by postponed reproductive maturation and enhanced survival under stressful conditions.

Death spiral   When individual organisms are dying, and are functionally different from those organisms in their cohort that are not dying.

Demographic heterogeneity, lifelong   When members of a cohort exhibit lifelong differences in their capacity to survive or to reproduce.

Demographic stochasticity   Variation that occurs in the reproduction of small populations due to variation in offspring production of individual females.

Demography   The study of the vital statistics of populations, such as age-specific birth and death rates.

Density dependence When the number of individuals in a given population affects demographic properties of that population, such as survival rates or fecundity.

Desiccation resistance The ability of organisms to survive conditions with limited water stores or accessibility, vid. Stress resistance.

Development The period between the onset of life and the onset of reproduction.

Dietary restriction An experimental procedure in which cohorts of animals are given reduced levels of nutrients.

Differential gene expression Differences in rates or kinds of transcription of mRNA sequences.

Diploid When all chromosomes are present in exactly two copies.

Directional selection When selection consistently favors an extreme phenotype.

Discrete generations When the parents of each generation never mate with individuals of other generations.

Disposable soma hypothesis The hypothesis that natural selection favors life cycles in which later somatic survival is sacrificed for increased early reproduction. An example of antagonistic pleiotropy.

Disruptive selection When selection consistently favors two or more contrasting phenotypes. For example, selection for both low and high body weights.

Divergence rate The number of genetic substitutions that have occurred during the divergence of two species, per unit time.

Divergence time The time since two species or lineages last shared a common ancestor.

DNA repair Replacement of nucleotides missing from a DNA molecule.

Dominance When a heterozygote's phenotype is not precisely intermediate between the phenotypes of its respective homozygotes; or, the quantitative deviation of a heterozygous phenotype from the mid-point between the homozygous phenotypes.

Dysdifferentiation hypothesis The explanation of aging as a byproduct of progressively dysfunctional regulation of somatic cell function.

Ectotherm Animals that maintain their body temperature with external sources of energy.

Effective population size (N_e) A measure of the number of individuals contributing offspring to the next generation, weighted by their net fertility.

Electrophoresis The separation of large molecules by moving them through charged gels.

Endotherm Animals that maintain their body temperature with internal sources of energy.

Epistasis When phenotypes that are determined by alleles at more than one locus are not predictable as additive combinations of the genotypes at individual loci; or, the quantitative deviation of phenotypes from such additive predictions.

Equilibrium population size The population size where births exactly equal deaths.

Error catastrophe A hypothesis that explains senescence as a result of positive feedback in the propagation of errors in the machinery of transcription or translation, leading to a collapse of cell function.

Eukaryote An organism with a nucleus distinct from the cytoplasm and other characteristic organelles, such as mitochondria or chloroplasts.

Euler-Lotka equation An equation that can be solved to determine the Malthusian parameter associated with a species, population, cohort, or genotype as a function of its age-specific survival and its age-specific fecundity.

Evolution A synonym for the Darwinian theory of life; or, change in the genetic composition of a population from one generation to the next.

Evolutionary heterogeneity fecundity model   Demographic models that incorporate a dying phase, distinct from both aging and late life.

Evolutionary history   The patterns of ancestry and selection that led to a species or population's current evolutionary state.

Exon   A part of a gene's DNA sequence that codes for the amino or ribonucleic acids of a gene product.

Exponential population growth   The growth of a population that follows from the assumption of a constant per capita net rate of reproduction.

Extinction   The dying off of an entire species, without direct descendants.

Fecundity   A female's output of eggs within a specified period of time.

Fertility   A female's output of viable offspring within a specific period of time.

Fission   The symmetrical division of a cell or a multicellular organism.

Fitness, Darwinian   The average reproduction of an individual or genotype, calibrated over a complete life cycle.

Fixation   When an allele rises to such a high frequency that all other alleles at its locus are eliminated from a population.

Force(s) of natural selection   The intensity of natural selection acting on life-history characters as a function of age.

Fragmentation   The production of viable offspring by breaking off pieces of the parental body.

Free radicals   Highly reactive molecules or atoms, usually in charged ionic form.

Gamete   A cell which can form a zygote by syngamy.

Gene expression   The extent to which a gene's DNA sequence is transcribed.

Generation time   The average length of time between the reproduction of the parents of each generation.

Genetic disease   A medical disorder caused by a genetic change at a single locus.

Genetic drift   Fluctuation in the frequency of an allele as a result of accidents of genetic segregation, reproduction, and survival.

Genetic engineering   The deliberate modification of the genome of an organism's cells, whether somatic or gametic, using molecular technology.

Genetic variation   Differences in genome composition between organisms in a population.

Genome   The complete set of DNAs transmitted during reproduction.

Genomics   Pertaining to the study of the entire genome, the comprehensive set of DNA from a given organism.

Genotype   The allelic contents of a locus, chromosomal region, chromosome, or entire genome.

Genotype-environment interaction   When the effect of particular alleles depends on the environment in which the organism lives.

Geriatrics   The medical specialty concerned with the treatment of elderly.

Germline engineering   Genetic engineering of the cells that will be used to make gametes.

Gerontology   The scientific study of aging in all organisms.

Gonad   An organ that produces gametes.

Group selection   Natural selection that arises from differences in the rates of extinction and colonization between groups that have different genetic compositions.

GWAS   Acronym for Genome-Wide Association Studies, research that correlates variation in SNPs with phenotypic variation, within or between populations.

Hamiltonian Theory Analysis of aging based on Hamilton's formulas for the forces of natural selection.

Haploid A full set of chromosomes without any duplicate chromosomes.

Hardy-Weinberg equilibrium The evolutionary genetic result that one generation of random mating in an infinite population leads to genotypic frequencies that are products of allele frequencies.

Hayflick Limit The maximum number of fissile divisions a normal somatic cell lineage can achieve in vitro.

Healthspan The duration of functional lifespan, as opposed to mere longevity.

Heat shock Physiological reaction to elevated ambient temperature.

Heritability The ratio of the genetic variance to the total phenotypic variance, when inheritance is strictly additive; with more complex patterns of inheritance, heritability is no longer given by this ratio.

Hermaphrodite An organism with both male and female functional sex organs.

Heterozygosity When a diploid locus has two different alleles; or, the fraction of the loci that are heterozygous.

Heterozygote superiority When the Darwinian fitness of a heterozygote is higher than that of homozygotes for its constituent alleles.

Homology When species have similar characteristics due to descent from a common ancestral species.

Homozygosity When a diploid locus has two copies of the same allele.

Huntington's disease A dominant genetic disease characterized by progressive loss of central nervous system function. Synonymous with Huntington's Chorea.

Hutchinson-Guilford syndrome A rare genetic disease that starts in childhood, featuring wrinkling, hair loss, and atherosclerosis; also known as childhood progeria.

Hybrid breakdown Reduced fitness of the offspring produced when the hybrids of two species mate with each other.

Hybrid dysgenesis Reduced fitness in the hybrids of old and new laboratory stocks of *Drosophila*, due to the proliferation of the P transposable element.

Hybrid vigor Increased fitness in offspring produced from crosses of differentiated stocks, breeds, varieties, or selected lines.

Hybridization When two species successfully mate and produce offspring.

Hybrids Individuals produced by the crossing of two species, two or more populations, or two or more differentiated laboratory stocks.

Immortality, biological The condition of roughly constant, positive, survival rates relative to age under good conditions; may not apply to all life-cycle stages.

Inbred line A lineage of experimental organisms that has been maintained with high levels of inbreeding for some time.

Inbreeding Mating of close relatives, which may be transitory or sustained.

Inbreeding depression Reduced fitness in an inbred organism; or, reductions in other functional characters in an inbred organism.

Incest avoidance When individuals who related avoid mating; or, when individuals who are raised together avoid mating.

Indeterminate growth Growth without a clear limit, characteristic of many tree and fish species, among other organisms.

Indy A supposed "longevity mutant," stands for "I'm not dead yet".

Intrinsic rate of increase   The parameter r of the logistic equation that determines the maximum rate of populations growth at low densities; also the Malthusian parameter in the Euler-Lotka equation.

Intron   DNA sequences that interrupt the coding sequences of genes; normally excised during transcription.

Isogamous   When species lacks obviously differentiated gametes.

Iteroparous   When life-histories involve repeated bouts of reproduction during adult life; in plants, perennial is a synonym.

Just-So stories   Explaining the features of organisms in terms of untested, or untestable, selection hypotheses.

Lamarckism   The doctrine that evolution occurs in independent lineages without branching and without natural selection, the direction of evolution arising from some kind of endogenous adaptive process.

Lansing Effect   When reproduction of individuals exclusively at later ages leads to accelerated aging.

Late life   The period of mortality-rate and fecundity stabilization after aging.

Life cycle   The sequence of birth, development, reproduction, and death that defines the biology of a species; or synonym for life history.

Life history   The quantitative features of the survival and reproduction of a species.

Life phase   A set of age-classes characterized by a pattern of mortality distinct from other life-phases. In outbred sexual populations, these might be development, aging, and late life.

Life table   A tabular summary of the qualitative life history of a species.

Lifelong demographic heterogeneity   See Demographic heterogeneity.

Lifespan   The longevity of individual organisms, or the maximum longevity observed for a species.

Linkage disequilibrium   When the frequencies of gametes are not given by the product of the frequencies of the alleles at multiple loci.

Linkage equilibrium   When the frequencies of gametes are given by the product of the frequencies of the alleles at individual loci.

Logistic population growth   A model of population growth that assumes per capita growth rates decline as a linear function of population size.

Malthusian parameter   A synonym for the intrinsic rate of increase; the rate of growth of a population when it achieves its stable distribution.

Maximum longevity   The greatest longevity reliability recorded for a member of a particular species.

Messenger RNA   The RNA transcribed from the DNA of a gene.

Metabolomics   The study of an extensive set of metabolites obtained from whole organisms or their specific tissues; usually focused on quantitative measures of the abundance of such metabolites.

Metazoan   A multicellular animal.

Mitochondrial DNA   The bacteria-like genome of the eukaryotic mitochondrium.

Muller's Ratchet   The progressive loss of individuals with low numbers of deleterious mutations due to a genetic drift in finite populations.

Mutagenesis   Process by which DNA is altered, creating a mutant gene.

Mutation accumulation   The evolutionary accumulation of alleles with deleterious effects that occur so late in adult life that they have little or no effect on Darwinian fitness.

Mutational load The reduction in average Darwinian fitness resulting from the accumulation of deleterious mutant variants in a population.

Natural selection The differential net reproduction of genetically distinct entities, whether mobile genetic elements, organisms, demes, or entire species.

Nematode Roundworms from the phylum Nematoda.

Neobalance model Evolutionary genetic theory that outbred Mendelian populations can harbor abundant standing genetic variation of functional significance due to some form of balancing selection.

Neoclassical model Evolutionary genetic theory where abundant segregating genetic variation genome-wide is explained exclusively as neutral or weakly deleterious alleles maintained by recurrent mutation.

Net reproductive rate The total reproductive output of an organism, discounted according to survival probabilities during juvenile and adult stages.

Neutral theory The theory that most molecular genetic variation is due to alleles that do not affect Darwinian fitness.

O populations Outbred populations of *D. melanogaster* selected for extended longevity by many generations of culture reproduction by older males and females exclusively.

Omic Pertaining to the large-scale study of omic data, e.g., genomics and transcriptomics.

Oncogene Genes that can mutate to alleles that foster carcinogenesis.

Oogenesis The production of eggs from germ cell lineages.

Opa1 A locus that can feature alleles that produce heart and eye defects.

Outbred When populations, breeds, or varieties are not systematically inbred; or, the population is not inbred and the population size is large.

Parthenogenesis The production of offspring without sex.

Perennials Plants that survive successive years. Some may be monocarps (semelparous) and flower once; others may be polycarps (iteroparous) and flower in successive seasons.

Phenotype A material attribute of an organism, excluding the genome.

Phenotypic selection Selection that depends on the phenotype of the organism.

Pleiotropy When allelic variation at a single locus has two or more phenotypic effects.

Polymorphism, phenotypic or genetic The presence of multiple distinct phenotypes in a population; or, the presence of multiple alleles at a genetic locus.

Poolseq A tool for next generation sequencing, where the genomes of individual organisms are physically "pooled" prior to shotgun sequencing.

Population A group of sexual organisms that interbreed frequently; or, a group of closely related asexual organisms that share a particular habitat.

Population genetics The study of the genetic composition of the individuals that together constitute a population.

Postponement of aging Shifting the onset of senescence towards a later age in a cohort or population.

Pseudogene A DNA sequence that is similar to a transcribed gene but is not itself transcribed.

Purifying selection When natural selection eliminates individuals who deviate markedly from the average phenotype of a species.

Quantitative character A biological character that can be measured numerically, such as body weight.

Quantitative genetics   The study of the variation and correlations among relatives for phenotypic characters, featuring the indirect inference of the genetic variation underlying phenotypic variation.

r- and K-selection   Natural selection under conditions of low and high crowding, respectively.

Rate-of-living hypothesis   The hypothesis that longevity varies inversely with metabolic expenditure.

Recessive   When an allele does not affect, or has less effect on, the phenotype compared to the allele(s) with which it is paired in heterozygotes.

Recombination   The re-assortment of alleles among gametes; may require physical breaking and re-joining of chromosomes when the alleles are located on the same chromosome; also occurs among alleles on separate chromosomes during meiosis.

Reproduction   Biological process of creating offspring can occur both sexually or asexually.

Reproductive costs   Reductions in survival or subsequent reproduction arising from investments in reproduction, whether reproduction itself or functional mobilization for reproduction.

Reproductive value   The expected contribution of an individual from a specific age group to the future reproduction of a population.

Selection differential   The phenotypic difference between a selected group and the population from which it was obtained.

Selection response   The phenotypic difference between the offspring of a selected group and the population from which the selected group was obtained.

Selective sweep   When a new mutant is selectively favored across all genotypes, thereby proceeding to fixation. This process purges genetic variation in the immediate vicinity of the mutation.

Self-fertilization, selfing   When an organism mates with itself to produce offspring.

Semelparous   When the life history of an organism has just one bout of reproduction.

Senescence   A synonym for aging, except in the case of botany where it may also refer to the deterioration and loss of deciduous leaves or flowers.

Sex ratio   The ratio between males and females in a population, expressed as a percentage or ratio.

Sexual reproduction   Reproduction in which gametes are produced by meiosis and these games then undergo syngamy to form zygotes.

Sexual selection   When natural selection acts on attributes that determine mating success.

Sib analysis   A quantitative genetic method for inferring the genetic and environmental components of variation underlying total phenotypic variation; the method uses patterns of variation and covariation among half-siblings, siblings, and unrelated individuals.

Single-Nucleotide Polymorphisms (SNP)   Variation at a site in the genome due to a single nucleotide difference.

Somatic mutation hypothesis   The hypothesis that aging arises physiologically from the accumulation of mutations in the genomes of somatic cells.

Stabilizing selection   Selection against phenotypic extremes.

Stable age-distribution   A property of age-structured populations where the proportions of individuals in each age class remain constant over time.

Starvation resistance   The ability of organisms to survive conditions with limited available calories, vid. Stress resistance.

Stress resistance   The ability of organisms to survive environmental stressors like starvation and desiccation.

Tinman   A Drosophila mutant allele that impairs heart development and subsequent function.

Transcription   The production of RNA sequences from a template of genomic DNA with the corresponding sequence of nucleic acids.

Transcriptomics   The study of the transcriptome, the comprehensive set of transcripts created from the genome.

Translation   The use of an RNA molecule to assemble a polypeptide based on the codon sequence of the RNA.

Transposable element   A DNA sequence that can make copies of itself that move to new locations in a host genome.

Variance   The average squared deviations from the mean of a random variable.

Variance, additive genetic   That part of the genetic variance which breeds true, that is, which causes predictable resemblance between relatives.

Variance, genetic   That part of the phenotypic variance which can be attributed to genetic causes, usually inferred from the resemblance of relatives.

Variance, phenotypic   The variance between organisms for a quantitative character.

Vegetative reproduction   Reproduction in plants and animals from somatic cells without formation of an egg or of gametes.

Viability   The probability of survival from zygote to adulthood.

Wear-and-tear   Mechanical or molecular damage caused by external injury or motor and other forms of activity.

Werner's Syndrome   A genetic disease that produces the appearance of accelerated aging, also known as adult progeria.

Yeast   A unicellular fungal species, in general; more specifically, the yeast species *Saccharmoyces cervisiae*.

Zygote   The cell stage at the start of organismal development, usually a diploid cell produced by syngamy of haploid egg and sperm.

Author Index

A

Abdel-Aal, Y., 202–203
Abugov, R., 102
Ackermann, M., 217
Akasaka, T., 252
Albert, P. S., 169–170
Albertini, R., 53
Alipaz, J. A., 174, 202, 222
Alla, R., 96
Allikian, M. J., 151
Anholt, R. R. H., 107
Archer, J. R., 154
Archer, M. A., 207–208
Aristotle, 7–8
Arking, R., 89, 92
Arking, R., 133–134, 142–143, 163, 208, 222
Arnold, K. R., 251–252
Austad, S. N., 154, 163–165, 197
Avise, J. C., 4
Ayala, F. J., 130–131, 150–151
Ayroles, J. F., 107
Ayyadevara, R., 143
Ayyadevara, S., 96, 143

B

Bacon, F., 12
Barbadilla, A., 107
Barker, D., 161
Barnes, L. L., 153, 173
Barrick, J. E., 221, 243–244
Barro, R., 1
Barrón, M., 107
Barrows, C. H., 40
Barter, T. T., 222–223, 227–230, 240–241, 244, 247–248, 251–252
Baselga, M., 164
Beard, R. E., 185
Beckman, K. A., 207–208
Bell, G., 16, 56, 110–111, 118
Bennett, A. F., 120–121, 244
Benzer, S., 194
Berg, P., 164
Bess, C., 107

Bidder, G. P., 43, 60, 173
Bierbaum, T. J., 130–131
Birchenall-Sparks, M. C., 153
Birse, R., 251–252
Blankenburg, K. P., 107
Bodmer, R., 251–252
Borash, D. J., 178
Bradić, M., 134
Bradley, T. J., 182, 208, 251–252
Brar, H., 150–151
Bray, N. L., 227
Broman, K. W., 239
Bross, T. G., 212
Buck, S. A., 89, 92, 133–134, 142–143, 163, 195, 208
Buetz, C., 218
Bui, L., 255–257
Burcombe, J. V., 123–124
Burke, M. K., 164, 193, 221–223, 227–229, 239–241, 243–244
Burke, M. K., 247–248, 251–252
Butler, D. G., 222
Butler, J. A., 153
Butlin, R. K., 114

C

Cabral, L. G., 1, 222–223, 227–229, 240–241, 243–245, 247–248, 251–252, 255–257
Campbell, T., 222
Canfield, A. L., 1
Carbone, M. A., 107, 222
Carey, J. R., 159–161, 185, 189–190, 212
Carnes, M. U., 222
Casillas, S., 107
Castellano, D., 107
Cervantes, B. G., 1
Chaboub, L., 107
Chang, J., 169–170, 197
Chang, P. L., 222, 227–228, 230, 240–241
Chao, J. U., 218
Chao, L., 218

Townsend, D. R., 159–160, 185,
189–190, 212
Tracey, M. L., 150
Tran, X., 202, 233–235
Travisano, M., 88, 92
Tse, S., 252
Tyler, R. H., 150–151

U
Ursúa, J., 1

V
Van Voorhies, W. A., 121, 170, 193–195,
197–198, 223–224, 248
Vaupel, J. W., 159–161, 185, 189–190,
201, 212, 239
Vertino, A., 143
Villeponteau, B., 244
Vu, L. N., 158, 165, 207

W
Wachter, K. W., 186
Walker, D. W., 198
Ward, S., 197–198
Wattiaux, J. M., 61, 66, 88–89
Weisman, G., 149–151
Weismann, A., 16, 23
Weitzel, A., 222
Weng, J., 1

Wessells, R. J., 252
Williams, C. G., 16, 36, 88, 129, 134, 161,
165–166, 255
Williamson, J. A., 73–76, 99
Wilson, E. O., 129–130
Wilson, J. B., 235–236
Wolverton, T., 212
Wong, B. D., 251–252
Wu, J. H., 143

X
Xiu, L., 189–190

Y
Yan, A., 251–252
Yasuda, S., 252
Yazdi, P. G., 201–202, 233, 248
Yee, K. J., 182
Yoon, S. H., 243–244
Yu, D. S., 243–244

Z
Zettler, E. E., 84, 106
Zhang, D., 239
Zhao, J. H., 239
Zhou, S., 222
Zhu, D., 107
Zwaan, B., 174

Index

Printed in the United States
by Baker & Taylor Publisher Services